Reclaiming the Knowledge Economy

"Going beyond critique to explore many inspiringly positive policy and political implications, this book offers a pioneering understanding of alternative agro-food networks as an emerging new form of "knowledge economy". Drawing on a diversity of theoretical traditions—from Marxist sociology to heterodox economics and science and technology studies—it offers an account of an in-depth multi-methods empirical study of some important agro-food initiatives in the UK. The resulting thoughtful and provocative call to re-claim ideas of the knowledge economy, holds deep cross-national relevance for contemporary food production, as well as repercussions across wider sectors."
—Professor Andy Stirling, *Science Policy Research Unit, University of Sussex, UK*

"Using the example of food networks and the roles of scientific laboratories, farmers and consumers, this important book makes a powerful argument for a collective re-thinking and reclaiming of the "knowledge economy", to create more inclusive economic systems based on a diversity of knowledge and experience."
—Dr Helen Wallace, *GeneWatch, UK*

"Critiques of "the (bio)knowledge economy" abound, as do critical exposures of the unjustness and unsustainability of conventional global agri-food systems. This book offers a new departure by developing a theoretically literate, empirically rich investigation of alternative agri-food networks, asking how their less studied, less formal but substantive knowledge-economy qualities might help in the urgently needed redesign and repurposing of dominant techno-capitalist agrifood systems, and the scientific R&D cultures subordinated to them."
—Professor Brian Wynne, *Lancaster University, UK*

Katerina Psarikidou

Reclaiming the Knowledge Economy

The Case of Alternative Agro-Food Networks

Katerina Psarikidou
University of Sussex
Brighton, UK

ISBN 978-981-16-6842-5   ISBN 978-981-16-6843-2 (eBook)
https://doi.org/10.1007/978-981-16-6843-2

© The Author(s), under exclusive licence to Springer Nature Singapore Pte Ltd. 2021
This work is subject to copyright. All rights are solely and exclusively licensed by the Publisher, whether the whole or part of the material is concerned, specifically the rights of translation, reprinting, reuse of illustrations, recitation, broadcasting, reproduction on microfilms or in any other physical way, and transmission or information storage and retrieval, electronic adaptation, computer software, or by similar or dissimilar methodology now known or hereafter developed.
The use of general descriptive names, registered names, trademarks, service marks, etc. in this publication does not imply, even in the absence of a specific statement, that such names are exempt from the relevant protective laws and regulations and therefore free for general use.
The publisher, the authors and the editors are safe to assume that the advice and information in this book are believed to be true and accurate at the date of publication. Neither the publisher nor the authors or the editors give a warranty, expressed or implied, with respect to the material contained herein or for any errors or omissions that may have been made. The publisher remains neutral with regard to jurisdictional claims in published maps and institutional affiliations.

Cover pattern © Melisa Hasan

This Palgrave Macmillan imprint is published by the registered company Springer Nature Singapore Pte Ltd.
The registered company address is: 152 Beach Road, #21-01/04 Gateway East, Singapore 189721, Singapore

In memory of my Father, Giorgos Psarikidis.

Preface and Acknowledgements

With this book, I express my sincere hope to contribute to the enactment of more inclusive research, policy and innovation agendas for agrifood sustainability. This is a strong commitment of mine, which has also been consolidated through my engagement in a number of agrifood research projects and, most importantly, my engagement and collaboration with a number of agrifood initiatives conducting important work on the ground. I am grateful to both of them, but especially to the latter for offering to me the unique opportunity to discuss, participate, learn and reflect, and therefore appreciate the plurality of knowledges and agrifood experts that are important for the production of new knowledge and innovation for agrifood sustainability. It is this particular message that I would also like to support as well as strongly communicate through this book and my attempt to reclaim the knowledge economy through AAFNs: underline the need for pluralising research, policy and innovation agendas by taking into consideration the knowledges, voices and standpoints of a broader spectrum of experts who are important in configuring as well as enacting today's knowledge economies.

I would therefore like to start by extending my sincere thanks to the alternative agrifood initiatives that I was honoured to have met through this research and for all I have learnt from them. I would like to gracefully thank all research participants for kindly providing their precious time and hospitality, and, above all, for being a great inspiration for this book and its argument.

I am privileged to be part of a wonderful community of enlightened academics and researchers that have provided me with the intellectual support and thought-provoking, stimulating discussions at different stages of my academic career around this topic. I am grateful to Science Policy Research Unit (SPRU) of the University of Sussex for providing me with such an intellectually stimulating environment to work in. I am grateful to both colleagues and students at SPRU for inspiring discussions through research and teaching.

I would like to extend my gratitude to my PhD supervisors back at Lancaster University, Professor Bron Szerszynski and Professor Larry Busch, for their generous input and invaluable conversations that have been foundational to my thinking and writing. I am grateful to my colleagues from the EC FAAN project and particularly Dr Helen Wallace from GeneWatch UK, Dr Sandra Karner from IFZ Austria and Dr Les Levidow from the Open University for stimulating discussions around local food networks. I am particularly grateful to GeneWatch UK, for which I am honoured to have worked for and learnt from.

I am also grateful for all I've learnt about agrifood research from more recent agrifood projects I have been involved in. I would like to particularly thank Professor Claire Waterton for working together on the HEFCE Agrifood Resilience programme, and Dr Elaine Swan and Professor Carol Wagstaff for our inspiring work around research co-production as part of the UKRI 'Co-producing healthy, sustainable food systems for disadvantaged communities'. My special thanks also go to a number of colleagues for their invaluable input and support at different stages of my research journey: Professor Brian Wynne, Dr Elaine Swan, Dr Marianna Cavada, Dr Manolis Papaioannou, Dr Ebru Thwaites, Professor Bulent Diken, Professor Andrew Sayer, Professor Bob Jessop, Professor John Urry, Dr Saurabh Arora, Dr Rachael Durrant, Dr Adrian Ely, Professor Andy Stirling.

Last, but definitely not least, I would like to thank my family for being an inspiration and motivation for me. I am grateful to my daughter Melitta Tyson and my partner Chris Tyson for their encouragement, understanding and support throughout the book-writing process, especially under the unprecedented circumstances of the Covid-19 pandemic. My gratitude to my parents, Soultana and Giorgos Psarikidis, and my brother Thanasis Psarikidis, for their continuous and sincere support and belief in me. My thanks to Gillian and Stephen Tyson and to Alexandra Petrescu for supporting our family.

Many thanks also to Palgrave for offering an intellectual home and communication platform for my work. My special thanks to the editor Josh Pitt for his support throughout the process.

I would like to devote this book to the memory of my Father, Giorgos Psarikidis, who passed away when I was working on the proofs of my book. I am grateful for all he has offered to me and I have learnt from him.

Brighton, UK   Katerina Psarikidou

Contents

1 Introduction: Bringing Together Alternative Agro-Food Networks and the Knowledge Economy: Why This Study, Why Now?   1

2 Re-inventing Capitalism Through the Knowledge Economy   19

3 Understanding the Ago-Food Economy as a Knowledge Economy   45

4 Researching Alternative Agro-Food Networks in the Aftermath of the Knowledge Economy   73

5 Re-Thinking the Knowledge Economy Through Alternative Agro-Food Networks   99

6 Conclusions: Reclaiming the Knowledge Economy   133

References   143

Index   165

CHAPTER 1

Introduction: Bringing Together Alternative Agro-Food Networks and the Knowledge Economy: Why This Study, Why Now?

A Few First Words

In recent years, a 'cognitive turn' has become important in the making of contemporary societies and economies. Especially, with the rise of the newly emerging industrial economies of the Global South, knowledge is increasingly configured as commodity (Jessop 2007), offering a new competitive economic advantage to the already 'developed' North. This vision is manifested in OECD and EU research and policy agendas that, as early as 2000, stated their aspiration to become 'the most competitive and dynamic knowledge-based economy in the world capable of sustainable economic growth with more and better jobs and greater social cohesion' (EC 2000). The concept of 'knowledge economy' presents the promise for a new 'post-industrial' economic development, which can be based not only on commodification of knowledge but also on life itself, as well as the emergence of new forms of 'productive' human *and* non-human labour. However, a narrow capitalocentric and technocentric understanding prevails through which 'knowledge economy' is primarily configured around the production of easily commodifiable techno-scientific knowledge that can be packaged and sold for profit (see Collins and Evans 2002).

The agro-food sector has been an interesting economic space for the materialisation of a knowledge-driven economic vision that can be based on the production of new techno-scientific knowledge about life and the subsequent commodification of both knowledge and life—for example, through the development of GM crop science and biofuels. However, it also provides an interesting economic space that manifests the plurality of knowledges and skills an economy can be based on. This is not only evident in the diversity of knowledges and knowledge production processes involved in the conventional agro-food sector, but also in the more recent agro-food economic developments that are usually configured as alternatives to the dominant food regime (McMichael 2009). However, less attention has been paid to the latter, usually also described as alternative agro-food networks (AAFNs).

This book aims to extend and reorient these debates, by specifically focusing on the case of 'alternative agro-food networks' and exploring their relationship to the 'knowledge-based economy' (KBE). By bringing into creative dialogue these two contemporary, and allegedly contradictory, economic developments, it is aimed to provide some critical sociological and political economic insights that can help us understand and critique the nature of contemporary dematerialised economies as well as re-claim and re-think the ways they are currently imagined and performed. An exploration of the Knowledge Economy (KBE) through the lens of AAFNs is pivotal towards such a direction and is unpacked through the chapters that follow.

However, what do we mean by AAFNs, and how do they relate to the KBE? Before delving into a deeper investigation of the complex interrelationship between the AAFNs and the KBE, the following sections of this introductory chapter provide a more detailed account of both concepts, also in ways that would help us realise the research need and timeliness for bringing these two areas of interest together.

ALTERNATIVE AGRO-FOOD NETWORKS: DEFINITIONS, THEORIES AND PERSPECTIVES

'Alternative agro-food networks' is an all-embracing term that describes networks of different organisations and actors—such as farmers and farm workers, retailers and consumers, environmentalists and food security advocates—that aim to pre-figure alternatives to the conventional agro-food system. Despite the apparent differences in their aims, strategies and approaches—which may also have worked against the construction of a unified political agenda—these actors variously coalesce towards the development of collective forms of resistance and counter-pressure to dominant food regimes. They thus aspire to develop a critique of corporate, intensive and de-personalised modes of agricultural production, distribution and consumption, but also, through that, contribute to more socio-economically just and environmentally responsible societies and futures.

Thus, 'alternative agro-food networks' is a concept widely used *and* debated in existing academic and non-academic circles—the latter mainly involving civil society and policy advocacy groups with an interest in sustainable farming and food. In academic circles, 'alternative agro-food networks' is used to describe the diversity of initiatives and networks of actors—for example, producers, retailers and consumers—that, by focusing on food, attempt to embody alternatives to the conventional industrialised, global agro-food system (e.g. Renting et al. 2003; Murdoch et al. 2000; Sonnino and Marsden 2006). Amongst civil society and policy advocacy circles, concepts such as 'food democracy', 'food sovereignty' and 'food justice' have been used to qualify the 'alternative' character of such networks: their possibility to configure more inclusive and socio-environmentally just agro-food systems that can be based on the voices, knowledges and needs of a wider spectrum of stakeholders, including small-scale farmers and other citizens (La Via Campesina 1996; People's Food Policy 2019). In particular, within 'modern geographies of food' (Whatmore and Thorne 1997), 'alternative agro-food networks' have been conceived as key expressions of a new 'post-productivist agro-food regime' (Whatmore et al. 2003; Ilbery and Bowler 1998) or a 'new rural development paradigm' (van der Ploeg et al. 2000; Marsden et al. 2000) with the potential to 're-spatialise' and 're-socialise' food by shortening the socio-spatial distance between producers, retailers and the end-users of the chain, as well as between people food and the environment (Renting et al. 2003).

This 'different' socio-spatial character of 'alternative agro-food networks' has been in the centre of academic work and attention. In their account of the contemporary agro-food system, Hendrickson and Heffernan underlined the ways 'space has been disconnected from place in the dominant food system' (2002, p. 349). Thus, alternative agro-food networks were viewed as a means of resistance to the agro-food distanciation and disconnection reproduced within the conventional agro-food system (Winter 2003, p. 508)—a possibility also embedded in the use of 'short food supply chains' (Renting et al. 2003; Marsden et al. 2000;

Ilbery and Maye 2005), 'local food networks/systems' (Morris and Buller 2003; Feenstra 1997; Henderson 1998; Lacy 2000; Hinrichs 2003; Ilbery et al. 2006) and 'alternative agro-food movements' (Henderson 2000; Buttel 1997; Hassanein 2003; Gottlieb 2001) as alternative concepts describing these initiatives. In this context, 'space' is not just a geographical territory reinforcing relations of physical proximity through a more direct 'reconnection' or 're-linking' between people and food (Hartwick 1998; Renting et al. 2003). It also acquires a wider socio-cultural meaning (Feagan 2007)—also depicted in the use of terms such as 'the quality turn' (Marsden 1998, 2004; Goodman 2003; Morris and Young 2000). The agro-food space is thus configured as a 'foodshed' (Kloppenburg et al. 1996) for the development of more hybrid socio-natural interactions and relations of intimacy between people, food and the environment, and diverse 'relations of regard' among proximal and distant, human and non-human others (Sage 2003; Hinrichs 2000; Psarikidou and Szerszynski 2012a, 2012b). By embedding a more qualitative sense of 'space', alternative agro-food networks position themselves as alternatives to the socio-spatially disconnecting nature of the contemporary agro-food system (Watts et al. 2005). Despite their embedded ambiguity, concepts such as 'local' and 'locality' also serve such purposes—by presenting another form of antinomy and resistance to the contemporary de-localised and socio-spatially distanciated agro-food system (Goodman and Redclift 1991; Raynolds 2000; McMichael 2000; Watts et al. 2005).

However, this very same concept of 'alternative agro-food networks' has also raised a debate with regard to their 'alternative character', and specifically their potential to constitute a 'real alternative', 'oppositional' agro-food movement. For example, Sonnino and Marsden (2006) questioned the possibility of drawing a clear line between the alternative and conventional food networks. Goodman and Goodman (2007) raised concerns about the alternative agro-food sector's possible appropriation—also called mainstreaming—by the conventional agro-food sector. Focusing specifically on the organic agro-food sector, scholars variously discussed the possible socio-economic inequalities within it, and questioned the political tractability of organic agriculture and its potential to create an alternative, socio-ecologically sustainable and just agro-food system (see also Guthman 2004; Raynolds 2004). The contested sustainability of these networks has also been analysed more broadly, underlining the inherent socio-economic inequalities, relations of power and social stigma that can be embedded within the alternative agro-food sector (see Allen et al. 1991; Ikerd 1993; Kloppenburg et al. 2000; Brown and Getz 2008; Allen and Wilson 2008; Getz et al. 2008; Psarikidou et al. 2019). For example, Watts et al. (2005) have analysed the increasing commodification potential of the alternative character—for example, organic, local—of the networks through mainstream certification and labelling schemes. Whereas Du Puis and Goodman (2005) have also talked about the possible 'defensive localist' risks embedded in 'the local', turning 'the alternative' into an elitist and reactionary attitude, a patriotic opposition and protectionist resistance to 'the other'.

This debate brings forth a question that Allen et al. (2003) have posed with regard to the 'actually existing' alternative character and oppositional potentiality of so-called 'alternative agro-food networks', a question which is also key in the investigation to follow: 'are AAFNs significantly oppositional or primarily alternative?' (p. 61). Are they aiming to change the structural relationships that characterised and supported industrial agriculture or are they mainly seeking 'innovative strategies to organise the production, exchange and consumption of food in alternative ways' (Allen et al. 2003, p. 73)?

These last questions provide some fertile ground for initiating our investigation of the 'alternative character' of AAFNs, especially in relation to the KBE. As discussed above, alternative agro-food networks are configured as 'alternatives' of opposition or resistance to the socio-economic and environmental challenges posed by the agri-industrial food model. Despite the increasingly important role of the KB(B)E in agriculture and food, existing work on 'alternative agro-food networks' has paid little attention to the ways AAFNs relate to the KB(B)E. This book aims to fill in this research gap, by specifically bringing into creative dialogue these two contemporary and, allegedly contradictory, developments: AAFNs and the KB(B)E. Following Allen et al.'s (2003) dual understanding of 'the alternative' within AAFNs, it is intended to engage us into a thought experiment through which we can delve deeper into the ways that AAFNs relate to the KB(B)E: not only by constructing 'an alternative to the Knowledge Economy', but also by constituting 'an alternative Knowledge Economy': an economy possible to develop its own innovative strategies to organise production, exchange and consumption of food differently.

Bringing Together Alternative Agro-Food Networks and Knowledge-Based Economies

Currently there is a growing academic literature in the fields of alternative agro-food networks and the Knowledge-Based (Bio)Economy (KBBE). However, such work still treats these fields as two distinct areas of research interest. This is the first book that brings together these two research areas, with a vision to problematise the concept of the 'knowledge economy' by specifically focusing on the potential of AAFNs to constitute a knowledge-based economy (KBE).

As discussed, studies of 'alternative agro-food networks' have significantly focused on identifying the 'alternative characteristics' of these networks. They have provided numerous conceptualisations of AAFNs, they have highlighted the great heterogeneity as well as the commonalities amongst the initiatives and actors involved, and unpacked the wider political economy of power and inequalities in which these networks can be embedded. Research has also looked into the policies, and other factors, both hindering and facilitating their future sustainability (e.g. Lang et al. 2009; Lang 2009; Marsden and Sonnino 2008; Morgan and Sonnino 2007). There is also a clear sustainability focus in existing work (Marsden et al. 1999; Kloppenburg et al. 2000; Ilbery and Maye 2005; Iles 2005; Pretty et al. 2005; Seyfang 2006), but also an investigation of the social and ethical relationalities embedded in such networks (e.g. Clarke et al. 2008; Goodman et al. 2010; Jackson et al. 2009; Lee 2000; Trentmann 2007).

Within such work, there is also a growing number of studies investigating the 'alternative' character of AAFNs, by specifically focusing on the knowledge patterns and knowledge production processes within them. For example, existing research highlights the interconnections between the contemporary European 'rural society' and 'knowledge society', and unpacking the complex knowledge production processes within the rural agro-food sector (see also Bruckmeier and Tovey 2008; Tovey 2008; Fonte 2008; Fonte and Grando 2006; Fonte and Papadopoulos 2010; Sielbert et al. 2008, etc.). These studies provide an initial and useful ground for this book's investigation of the diverse knowledge patterns and production processes within AAFNs. However, most of this work has primarily focussed on rural alternative agro-food systems. Also, it does not delve in an investigation that can unravel the possible 'knowledge economic' aspects of AAFNs, and, therefore, the ways in which AAFNs' 'alternative character' can be configured in relation to the KB(B)E. In their work, Morgan and Murdoch have been the first attempting to unpack the significance of knowledge in the configuration of alternative agro-food economy (see Morgan and Murdoch 2000). Thus, despite their primary and sole focus on the organic production methods as their case of investigation, their comparison of the knowledge dynamics within organic and conventional agricultural systems provides some useful theoretical insights and inspiration for the analysis in this book.

A similar observation can also be made about current analyses and studies of the knowledge economy (KBE). A lot of existing studies explore the knowledge-based economy (KBE) as a neoliberal imaginary and a techno-knowledge fix. They have critically engaged with the KBE as applied to various economic sectors (see Walby et al. 2019), including the agro-food sector. They paid particular attention to the role of technoscience in the transformation of various forms of life— for example, biomedicine—and offered critical insights on the role of such knowledge systems in the capitalisation and commodification of both knowledge and life (e.g. Sunder Rajan 2006; Cooper 2008). With regard to agriculture, along similar lines, much attention has been paid to the critique of techno-science in shaping agro-food research and innovation, and therefore the agrifood sector itself (e.g. Wynne 2005, 2007; Levidow and Carr 2007; Levidow and Boschert 2008; Reynolds and Szerszynski 2010). However, very little attention has been paid to the KBE per se as a powerful lens shaping the agricultural sector (e.g. Ponte 2009; Birch et al. 2010; Wallace 2010). Also, most of these studies primarily focused on developing a critique of the role of KBE in shaping modern agro-food spaces, thus ignoring the possibilities that can be opened up around alternatives. Thus, despite their acknowledgement of an urgency to develop alternative strategies for sustainable agriculture and food, little discussion is usually undertaken with regard to the existing or future alternatives and their possibility to construct a powerful 'alternative to KBE' or even an 'alternative KBE'. On the contrary, in most of existing work, the alternatives—either in the agro-food sector or any other sector shaped around the KBE narrative and vision—appear at the end of the analyses, as a speculative future solution that is worth being further researched, but usually remains under- or un-researched.

Therefore, there is still very little research on the alternatives to the KB(B)E and the ways that they can constitute a powerful alternative able to relate to, compete with or even transform the dominant KB(B)E vision. And, this book is especially developed to respond to this research need, specifically by setting this last point as the starting point and key premise for investigation. Of course, this project also builds on the limited but rather inspiring academic work in this field. For example, in his 'European quality agriculture as an alternative bio-economy', Les Levidow (2008) was the first to bring these two research areas together. His study primarily focused on unpacking the 'alternative as oppositional' character of alternative agro-food networks, by specifically studying agro-ecology as the alternative method of production. Agro-ecology was also the focus of Levidow, Birch and colleagues' (2012a, 2012b) follow-up studies which explored contending visions for the bioeconomy. Marsden's work on the 'eco-economy' (see Marsden and Farioli 2015) also primarily focuses on agricultural production and, more specifically, agro-ecology, however, paying less attention to the complex knowledge production processes involved in the wider supply chain. Also, by using the term 'eco-economy', like in Levidow's work, agro-ecology is conceived as oppositional to the KBE rather than an alternative framework through which we can re-think KBE. This book aims to extend this research, specifically exploring the potential of alternative agro-food networks of production, distribution and consumption to constitute not only an 'alternative to the KBE', but also an 'alternative KBE'. In doing so, it also aims to move beyond a macro level of analysis and a specific focus on agro-ecology, and pay particular attention to the knowledge practices and the discursive narratives of those on-the-ground agrifood innovators that are involved in alternative agrifood (knowledge) practices from farm to fork. Following Allen et al.'s (2003) conceptualisation of 'the alternative', it is intended to move beyond approaching AAFNs as only 'oppositional' to the dominant techno-scientific model of agrifood innovation and the KBBE, but also exploring the 'innovative strategies [through which they] organise the production, exchange and consumption of food in alternative ways' (p. 73) and can therefore help configure them as an alternative KBBE.

By doing so, this book is the first sociological study that goes beyond a critique of the current manifestation of the 'knowledge economy' in the field (e.g. GM food; biofuels) and focuses on answering the usually unanswered question around 'the alternatives' and the ways they can constitute not only an 'alternative to' the 'knowledge economy', but also an 'alternative' 'knowledge economy'. It is the first sociological study that aims to re-claim the concept of the 'knowledge economy' by providing a detailed analysis of the ways other economic developments, such as this of alternative agro-food networks, can constitute a 'knowledge economy'. It thus provides critical sociological and political economic insights that help problematise the 'knowledge economy' as a master economic narrative (e.g. at the OECD and EC levels), challenge its dominant capitalocentric and technocentric understandings in current research and policy agendas and reconceptualise it through the study of its alternative. It thus supports an alternative framework for conceptualising the knowledge economy through the lens of alternative agro-food networks and discusses the performative effect of such a possibility for the formulation of future policy and practice, research and innovation strategies.

Through this detailed theoretical and empirical analysis, this book also offers a timely new perspective on the role of knowledge in re-inventing contemporary capitalism through the dematerialisation of its contents (e.g. labour, commodity, life), but also in challenging contemporary capitalism through a re-conceptualisation of its contents (e.g. the concept of 'knowledge economy' itself). It is intended to provide a useful intellectual context for researchers interested in re-claiming the concept of 'knowledge economy' at a wider field of applications (e.g. biomedicine; energy; ICT). And, it is envisioned to constitute an inspirational piece of work for agrifood stakeholders and policy-makers, which will contribute to more inclusive processes and frameworks for future research, innovation and policy agendas that can be based on the voices and knowledges of a wider spectrum of stakeholders.

A Snapshot of the Book

The first chapter to follow argues that understanding the Knowledge Economy is key to understanding processes of re-inventing capitalism in contemporary societies. Drawing on the theoretical traditions of Marxist Sociology and Political Philosophy (e.g. Schumpeter 1942; Hardt 1999; Hardt and Negri 2004; Lazzarato 1996; Jessop 2007; Merton 1948), it is thus aimed to offer a better understanding of the KBE by contextualising it in the wider process of dematerialisation of contemporary capitalist economies. It discusses the emerging processes of transformation of knowledge into a commodity and labour into 'immaterial labour', and underlines their centrality in the introduction of a new phase of an increasingly dematerialised capitalist economy. It then analyses the centrality of the KBE in such processes. It outlines the diverse existing conceptualisations and interpretations of the KBE, including the concept of the Knowledge-based *Bio*-Economy (KBBE) and the significance of 'bio' in re-inventing capitalism in ways that can go beyond the ecological contradictions of previous economic forms (see Brennan 2000; Cooper 2008; Sunder Rajan 2006). It then focuses on the example of the agro-food sector. It introduces the significance of promise in understanding as well as performing such dematerialised forms of capitalism. It thus suggests approaching the KBE as an economic imaginary that becomes a self-fulfilling prophecy (Jessop 2008; Merton 1948)—that is, an economic imaginary that, in the promise of a certain future, makes certain presents and socio-economic realities possible.

The next chapter unpacks the ways in which an agro-food economy can be understood as a knowledge economy. It compares the different knowledge forms and production processes involved in both the knowledge economy and the agro-food economy and identifies the common threads, which could encourage us to approach an agro-food economy as a knowledge economy. Through such analysis, it is also intended to introduce us to the particular elements—for example, types and patterns of knowledge and conversion processes—that will help investigate the knowledge-economic aspects of alternative agro-food networks in the chapter to follow. More specifically, it starts by providing evidence of the significance of both explicit and tacit forms of knowledge as well as complex knowledge production and conversion processes for the production of innovation. It then also makes a similar argument for the knowledge production processes within the agro-food sector. Building on sociological, geographical and political economic studies of agriculture and food, it underlines the significance of both explicit and tacit forms of knowledge, as well as complex knowledge production and conversion processes within the agro-food sector (Stuiver et al. 2004; Funtowicz and Ravetz 1994a, 1994b; Nygren 1999; Anteweiler 2004). By specifically focusing on the case of organic agriculture, it also sets the ground for understanding the alternative agro-food economy as a knowledge economy, a call which is set out in the last section of this chapter and provides the fertile ground for further investigation in the chapters to follow.

Before moving to the investigation of alternative agro-food networks as an alternative knowledge economy, Chap. 4 is intended to offer an overview of the fieldwork: the research methods and case studies used. In doing so, it is also aimed to provide justification for its geographical focus on the UK, a country with a prominent knowledge-based economy, also with significant implications in the agro-food sector. In doing so, it also aims to provide justification of the spatio-temporal frameworks of this study. It provides details of the geographies and the networks under investigation, as well as of the particular types, structural morphology and characteristics of the agro-food initiatives, also by situating them in the broader theoretical-methological framework and its instrumental role in not only researching but also contributing to the enactment of those alternative agro-food knowledge economies.

Building on all above, the fifth chapter conceptualises alternative agro-food networks as an alternative knowledge economy. Drawing on fieldwork with urban and rural agro-food initiatives in the UK, it presents the ways scientific and local, explicit and tacit forms of knowledge get conceptualised, perceived and performed through engaged actors' discourses and practices. The chapter is mainly divided into three distinct empirical sections that provide detailed analysis of the knowledge patterns and production processes within the alternative agro-food sites of production, distribution and consumption. The first section focuses on the site of agricultural production and revolves around the investigation of the complex ways scientific knowledge is both perceived and practiced by AAFN stakeholders. Focusing on distribution and consumption, the other two sections re-iterate this point, by providing detailed evidence of the fruitful dialogue, recombination and adaptation of different forms of knowledge within the alternative agro-food sector. This leads us to the final section of this chapter, which discusses the ways alternative agro-food networks constitute a knowledge economy. It identifies the characteristics through which an alternative agro-food economy resembles a knowledge economy, but also the ways it is different from its mainstream capitalocentric understanding. In other words, it discusses the ways an alternative agro-food economy also constitutes an *alternative* knowledge economy, in the way that it remains an integral part of it, while obtaining some qualitative characteristics which can significantly differentiate it from the dominant conceptualisation and understanding of it. In doing so, it sets the grounds for reclaiming the knowledge economy—an analysis that is further elaborated in the final chapter of this book.

Drawing together the analyses of the book, the last chapter aims to reclaim the concept of the 'knowledge economy'. Following Gibson-Graham's (1996, 2006) post-capitalist iceberg economy, it encourages us to think of the 'knowledge economy' as a diverse economic landscape of cohabitation and contestation between different 'knowledge economic' forms that are all equally important in the configuration of the 'knowledge economy'. It suggests moving beyond a singular and narrow capitalocentric and technocentric approach to the knowledge economy, and consider these other 'knowledge economies' that currently exist but remain silenced from today's research and policy agendas. The book finishes by calling for a reconceptualisation of the 'knowledge economy' in ways that would lead to the inclusion of a wider spectrum of stakeholders and knowledges in the production of new knowledge for agriculture and food. In doing so, it also calls for the reconceptualisation of the knowledge economy beyond the context of the agro-food sector and proposes the establishment of more inclusive processes of research and innovation also in other economic sectors that are currently described or framed as 'knowledge economic' sectors.

REFERENCES

Allen P., FitzSimmons M., Goodman M. and Warner K. (2003) Shifting plates in the agrifood landscape: the tectonics of alternative agrifood initiatives in California, *Journal of Rural Studies* 19(1): 61–75.

Allen, P., Van Dusen, D., Lundy, J. and Gliessman, S. (1991) Integrating social, environmental, and economic issues in sustainable agriculture, *American Journal of Alternative Agriculture* 6(1): 34–39.

Allen, P. and Wilson, A.B. (2008) Agri-food inequalities: globalization and localization, *Development* 51(4): 534–540.

Anteweiler, C. (2004) Local knowledge, Theory and Methods: an urban model from Indonesia. In Bicker, A., Sillitoe, P. and Pottier, J. (eds.) Investigating Local Knowledge. New Directions, New Approaches. Ashgate: pp. 1–34.

Birch, K, Levidow, L. and Papaioannou, T. (2010) Sustainable Capital? The Neoliberalisation of Nature in the European Knowledge-based Bio-economy, *Sustainability* 2: 2898–2918.

Brennan, T. (2000) *Exhausting Modernity: Grounds for a new economy*. London and New York: Routledge.

Brown, S. and Getz, C. (2008) Towards domestic fair trade? Farm labor, food localism, and the 'family scale' farm, *GeoJournal* 739(1):11–22.

Bruckmeier, K. and Tovey, H. (2008) Knowledge in sustainable rural development: from forms of knowledge to knowledge processes, *Sociologia Ruralis* 48(3): 313–329.

Buttel, F. (1997). Some Observations on Agro-Food Change and the Future of Agricultural Sustainability Movements. In Goodman, D. and Watts, M. (Eds.) *Globalising Food: Agrarian Questions and Global Restructuring*. London: Routledge: pp. 344–365.

Clarke, N., Cloke, P. Barnett, C. and Malpass, A. (2008) The spaces and ethics of organic food, *Journal of Rural Studies* 24:219–230.

Collins, H.M. and Evans, R. (2002) The Third Wave of Science Studies: Studies of Expertise and Experience, *Social Studies of Science* 32(2):235–296.

Cooper, M. (2008) *Life as Surplus: Biotechnology and Capitalism in the Neoliberal Era*. Seattle: University of Washington Press.

Du Puis, E.M. and Goodman, D. (2005) Should we got home to eat? Towards a reflexive politics of localism, *Journal of Rural Studies* 21:359–71.

EC (2000) *The Lisbon European Council—An Agenda of Economic and Social Renewal for Europe*. Brussels: The European Commission, DOC/00/7.

Feagan, R. (2007) The place of food: Mapping out the 'local' in local food systems, *Progress in Human Geography* 31(1):23–42.

Feenstra, G. (1997) Local food systems and sustainable communities *American Journal of Alternative Agriculture* 12:28–36.

Fonte, M. (2008) Knowledge, Food and Place. A Way of Producing, a way of knowing, *Sociologia Ruralis* 48(3): 200–222.

Fonte, M. and Grando, S. (2006) A Local Habitation and a Name: Local Food and Knowledge Dynamics in Sustainable Rural Development, *CORASON project*. Available online at www.corason.hu.

Fonte, M. and Papadopoulos, A. (2010) *Naming Food After Places: Food Relocalisation and Knowledge Dynamics in Rural Development*. Ashgate.

Funtowicz, S.O. and Ravetz, J.R. (1994a) Uncertainty, Complexity and Post-Normal Science. *Environmental Toxicology and Chemistry* 13(12): 1981–1984.

Funtowicz, S.O. and Ravetz, J.R. (1994b) Emergent Complex Systems. *Futures* 26(6): 568–582.

Getz, C., Brown, S. and Shreck, A. (2008) Class Politics and Agricultural Exceptionalism in California's Organic Agriculture Movement. *Politics Society* 36(4): 478–507.

Gibson-Graham, J.K. (1996) *The End of Capitalism (As We Knew It): A feminist Critique of Political Economy*. Oxford UK and Cambridge USA: Blackwell Publishers.

Gibson-Graham, J. K. (2006) *A Postcapitalist Politics*. Minneapolis, MN: University of Minnesota Press.

Goodman, D. (2003) The quality 'turn' and alternative food practices: reflections and agenda. *Journal of Rural Studies* 19:1–7.

Goodman, D. and Goodman, M. (2007) Localism, Livelihoods and the 'Post-Organic': Changing Perspectives on Alternative Food Networks in the United States. In Maye, D, Holloway, L. and Kneafsy, M. (eds.) *Alternative Food Geographies Representation and Practice*. Elsevier.

Goodman, D. and Redclift, M. (1991) *Refashioning Nature*. London: Routledge.

Goodman, M.K., Maye, D. and Holloway, L. (2010) Ethical foodscapes?: Premises, promises, and possibilities, *Environment and Planning A* 42(8): 1782–1796.

Gottlieb, R. (2001) Environmentalism Unbound: Exploring new pathways for change. Cambridge, MA: The MIT Press.
Guthman, J. (2004) The trouble with 'organic lite' in California: a rejoinder to the 'conventionalisation' debate, *Sociologia Ruralis* 44(3): 301–316.
Hardt, M. (1999) Affective Labour, *Boundary 2* 26(2): 89–100.
Hardt, M. and Negri, A. (2004) *Multitude: War and Democracy at the Age of Empire*. New York: The Penguin Press.
Hartwick, E. (1998) Geographies of consumption: a commodity-chain approach, *Environment and Planning D: Society and Space* 16: 423–437.
Hassanein, N. (2003) Practicing Food democracy: a pragmatic politics of transformation, *Journal of Rural Studies* 19:77–86.
Henderson, E. (1998) Rebuilding local food systems from the grassroots up, *The Monthly Review: An Independent Socialist Magazine* 50:112–124.
Henderson, E. (2000) Rebuilding local food systems from the grassroots up. In Magdoff, F., Bellamy Foster, J. and Buttel, F.H. (Eds.) *Hungry for Profit: The agribusiness Threat to Farmers, Food and the Environment*. New York: Monthly Review Press: pp. 175–188.
Hendrickson, M. and Hefferman, W. (2002) Opening spaces through relocalization: locating potential resistance in the weaknesses of the global food system, *Sociologia Ruralis* 42:347–69.
Hinrichs, C. (2000) Embeddedness and local food systems: notes on two types of direct agricultural market, *Journal of Rural Studies* 16: 295–303.
Hinrichs, C. (2003) The practice and politics of food system localization, *Journal of Rural Studies* 16:33–45.
Ikerd, J. (1993) Two related but distinctly different concepts: organic farming and sustainable agriculture, *Small Farm Today* 10(1):30–31.
Ilbery, B. and Bowler, I. (1998) From agricultural productivism to post-productivism. In B. Ilbery (ed.) *The Geography of Rural Change*. London: Longman.
Ilbery, B. and Maye, D. (2005) Alternative (shorter) food supply chains and specialist livestock products in Scottish-English borders, *Environment and Planning A* 37(4):823–844.
Ilbery, B., Watts, D., Simpson, S., Gilg, A. and Little, J. (2006) Mapping local foods: evidence from two English Regions, *British Food Journal* 108(3):213–225.
Iles, A. (2005) Learning in sustainable agriculture: food miles and missing objects, *Environmental Values* 14(2): 63–183.
Jackson, P. Ward, N. and Russel, P. (2009) Moral economies of food and geographies of responsibility, *Transactions of the Institute of British Geographers* 34:12–24.
Jessop, B. (2007) Knowledge as a fictitious commodity : insights and limits of a Polanyian perspective. In: Bugra, Ayse and Agartan, Kaan, (eds.) *Reading Karl Polanyi for the twenty-first century: market economy as political project*. Palgrave, Basingstoke.

Jessop, B. (2008) *State power: a strategic-relational approach.* Cambridge: Polity Press.

Kloppenburg, J., Henrickson, J. and Stevenson, G.W. (1996) Coming in to the foodshed, *Agriculture and Human Values* 13:33–42.

Kloppenburg, J., Lezberg, S., De Master, K., Stevenson, G.W. and Hendrickson, J. (2000) Tasting Food, tasting sustainability: Defining the Attributes of an alternative food system with competent, ordinary people, *Human Organisation* 59 (2): 177–186.

Via Campesina (1996) *Tlaxcala Declaration of the Via Campesina.* April 1996. Available online at http://www.viacampesina.org/en/index.php?option=com_content&view=article&id=445:ii-international-conference-of-the-via-campesina-tlaxcala-mexico-april-18-21&catid=32:2-tlaxcala-1996&Itemid=48.

Lacy, W. (2000) Empowering Communities through public work, science and local food systems: revisiting democracy and globalisation, *Rural Sociology* 65:3–26.

Lang, T. (2009) Re-shaping the food system for ecological public health, *Journal of Hunger and Environmental Nutrition* 4(3/4): 315–335.

Lang, T., Barling, D. and Caraher, M. (2009) *Food Policy: Integrating Health, Environment and Society.* Oxford: Oxford University Press.

Lazzarato, M. (1996) Immaterial Labor. In Hardt, M. and Virno, P. (eds.) *Radical Thought in Italy: A Potential Politics,* Minneapolis: University of Minnesota: pp. 133–50.

Lee, R. (2000) Shelter from the storm? Geographies of regard in the worlds of horticultural consumption and production, *Geoforum* 31:137–157.

Levidow, L. (2008) European quality agriculture as an alternative bio-economy. In Guido Ruivenkamp, Shuji Hisano and Joost Jongerden (Eds.) *Reconstructing Biotechnologies: Critical Social Analyses.* Wageningen Academic: pp. 185–205.

Levidow, L. and Boschert, K. (2008) Coexistence or contradiction? GM crops versus alternative agricultures in Europe, *Geoforum* 39(1): 174–190.

Levidow, L. and Carr, S. (2007) GM crops on trial: technological development as a real-world experiment, *Futures* 39(4): 408–431.

Levidow, L., Birch, K., Papaioannou, T. 2012a. Divergent paradigms of European Agro-Food Innovation: the Knowledge-based Bioeconomy (KBBE) as an R&D agenda, *Science, Technology and Human Values,* 38(1): 94–125.

Levidow, L., Birch, K. and Papaioannou, T. 2012b. EU agri-innovation policy: two contending visions of the bio-economy, *Critical Policy Studies* 6(1): 40–65.

Marsden, T.K. (1998) New rural territories: Regulating the differentiated rural spaces, *Journal of Rural Studies* 14(1):107–117.

Marsden, T.K. (2004) The quest for ecological modernisation: re-spacing rural development and agri-food studies, *Sociologia Ruralis* 44:129–146.

Marsden, T.K. and Sonnino, R. (2008) Rural development and the regional state: Denying multifunctional agriculture in the UK, *Journal of Rural Studies* 24:422–431.

Marsden, T. and Farioli, F. (2015) Natural Powers: From the Bio-economy to the Eco-economy and sustainable Place-making, *Sustainability Science* 10: 331–344.

Marsden, T., Murdoch, J. and Morgan, K. (1999) Sustainable agriculture, food supply chains and regional development: editorial introduction, *International Planning Studies* 4(3):295–301.

Marsden, T.K., Banks, J. and Bristow, G. (2000) Food supply chain approaches: exploring their role in rural development, Sociologia Ruralis 40:424–438.

McMichael, P. (2000) The power of food. Agriculture and Human Values 17:21–33.

McMichael, P. (2009) A Food Regime Genealogy, *The Journal of Peasant Studies* 36(1): 139–169.

Merton, R.K. (1948) The Self-Fulfilling Prophecy, *The Antioch Review* 8(2): 193–210.

Morgan, K. and Murdoch, J. (2000) Organic versus Conventional Agriculture: Knowledge, Power and Innovation in the Food Chain, *Geoforum* 31:159–173.

Morgan, K. J. and Sonnino, R. (2007) Empowering Consumers: The Creative Procurement of School Meals in Italy and the UK, *International Journal of Consumer Studies* 31(1):19–25.

Morris, C. and Buller, H. (2003) The local food sector: A preliminary assessment of its form and impact in Gloucestershire, *British Food Journal* 105(8):559–566.

Morris, C. and Young, C. (2000) 'Seed to shelf', 'teat to table', 'barley to beer' and 'womb to tomb': discourses of food quality and quality assurance schemes in the UK, *Journal of Rural Studies* 16:103–115.

Murdoch, J., Marsden, T. and Banks, J. (2000) Quality, nature and embeddedness: some theoretical considerations in the context of the food sector, *Economic Geography* 76(2):107–125.

Nygren, A. (1999) Local Knowledge in the Environment_Development Discourse: From dichotomies to situated knowledges, *Critique of Anthropology* 19:267–288.

People's Food Policy (2019) *A People's Food Policy: Transforming Our Food System*. Available Online at www.peoplesfoodpolicy.org

Ploeg, J.D. van der, Renting, H., Brunori, G., Knickel, K. Mannion, J. Marsden, T.K., de Roest, K., Sevilla-Guzman, E. and Ventura, F. (2000) Rural development: from practices and policies towards theory. *Sociologia Ruralis* 40(4):391–408.

Ponte, S. (2009) From Fishery to Fork: Food Safety and Sustainability in the Knowledge-Based Bioeconomy, *Science as Culture* 18(4):483–495.

Pretty, J., Ball, A., Lang, T. and Morison, J. (2005) Farm costs and food miles: an assessment of the full cost of the weekly UK food basket, *Food Policy* 30(1):1–19.

Psarikidou, K. and Szerszynski, B. (2012a) Growing the Social: Alternative Agro-Food Networks and Social Sustainability in the Urban Ethical Foodscape, *Sustainability: Science, Practice and Policy* 8(1): 30–39.

Psarikidou, K. and Szerszynski, B. (2012b) The Moral Economy of Civic Food Networks in Manchester, *International Journal for the Sociology of Agriculture and Food* 19(3): 309–327.

Psarikidou, K., Kaloudis, H., Fielden, A. and Reynolds, C. (2019) Local Food Hubs in deprived areas: De-stigmatising Food Poverty? *Local Environment: The International Journal for Justice and Sustainability* 24(6): 525–538.

Raynolds, L. (2000) Re-embedding global agriculture: the international organic and fair trade movements. *Agriculture and Human Values* 17:297–309.

Raynolds, L.T. (2004) The Globalisation of Organic Agro-Food Networks, *World development*. 32(5):725–743.

Renting, H., Marsden, T. and Banks, J. (2003) Understanding alternative food networks: exploring the role of short food supply chains in rural development, *Environment and Planning A* 35(3): 393–411.

Reynolds, L. and Szerszynski, B. (2010) Contested agro-technological futures: the GMO and the construction of European space. In Robbins, P. and Huzair, F. (Eds.) *Transitioning the Life Sciences*, Springer.

Sage, C. (2003) Social embeddedness and relations of regard: alternative 'good food' networks in south-west Ireland. *Journal of Rural Studies* 19:47–60.

Schumpeter, J. (1942) *Capitalism, Socialism and Democracy*. New York: Harper.

Seyfang, G. (2006) Ecological citizenship and sustainable consumption: examining local organic food networks, *Journal of Rural Studies* 22(4):383–395.

Sielbert, R., Larchewski, L. and Dosch, A. (2008) Knowledge dynamics in valorising local nature, *Sociologia Ruralis* 48(3): 223–240.

Sonnino, R. and Marsden, T.K. (2006) Beyond the Divide: Rethinking Relations Between Alternative and Conventional Food Networks in Europe, *Journal of Economic Geography* 6(2): 181–199.

Stuiver, M., Leeuwis, C. and van der Ploeg, J.D. (2004)The Power of Experience: Farmers' Knowledge and Sustainable Innovations in Agriculture. In J.S.C. Winskerke and J.D. van der Ploeg (eds.) *Seeds of Transition: Essays on novelty production, niches and regimes in agriculture*. The Netherlands: Koninkijke van Gorcum BV.

Sunder Rajan, K. (2006) *Biocapital: The Constitution of Postgenomic Life*. Durham and London: Duke University Press.

Tovey, H. (2008) Introduction: Rural Sustainable Development in a Knowledge Society Era, *Sociologia Ruralis* 48(3):185–199.

Trentmann, F. (2007) Before 'fair trade': empire, free trade, and the moral economies of food in the modern world, *Environment and Planning D: Society and Space* 25:1079–1102.

Walby, S., Gottfried, H., Gotschall, K. and Osawa, M. (eds.) (2019) *Gendering the Knowledge Economy: Comparative Perspectives*. Palgrave MacMillan.

Wallace, H. (2010) Bioscience for Life? Who Decides what research is done in health and agriculture? Genewatch UK Report. March 2010.

Watts, D.C.H., Ilbery, B. and Maye, D. (2005) Making reconnections in agro-food geography: alternative systems of food provision, *Progress in Human Geography* 29(1): 22–40.

Whatmore, S. and Thorne, L. (1997) Nourishing networks: alternative geographies of food. In Goodman, D. and M. Watts (Eds.) *Postindustrial Natures: Culture, Economy and Consumption of Food*. London, Routledge: pp. 287–304.

Whatmore, S., Stassart, P. and Renting, H. (2003) What's alternative about alternative food networks? *Environment and Planning A* 35: 389–391.

Winter, M. (2003) The policy impact of the foot and mouth epidemic, *Political Quarterly* 74(1): 47–56.

Wynne, B. (2005) Risky Delusions: How GM Science has imagined—and provoked—its publics. In Taylor, I. and K. Barrett (Eds.) *Genetically Engineered Crops: Decision-making under Uncertainty*. Canada, UBC Press.

Wynne, B. (2007) Risky Delusions: Misunderstanding Science and Misperforming Publics in the GE Crops Issue. In I.E.P. Taylor (ed.) *Genetically Engineered Crops: Interim Policies, Uncertain Legislation*. Haworth Press.

CHAPTER 2

Re-inventing Capitalism Through the Knowledge Economy

KNOWLEDGE AND LABOUR IN CONTEMPORARY ECONOMIES

In recent decades, knowledge has become increasingly contextualised, shaped and organised around various and complex social and economic interests, challenges and crises contemporary societies are called to face. As far back as the early stages of the transition to post-industrial societies, Daniel Bell predicted a transition from an 'economising' to a 'socialising logic', which could enable the subordination of economic activity to a democratically controlled social engineering. However, with the advent of a new stage of informational capitalism, such considerations seemed to have failed, situating new services based on the manipulation and communication of knowledge and information at the heart of economic relations of production and visions of productivity.

The importance of knowledge for future economic growth is not a new idea. Since the early stages of industrialisation, knowledge was acknowledged as a key aspect in the economic organisation of social life. Karl Marx wrote extensively on the role of knowledge in the construction of the social relations of production. In his discussion of the extensive mechanisation of industrial economies, he underlined the significance of the *general intellect*,[1] also signalled by the transformation of knowledge from an

---

[1] According to Marx, the general intellect—that is, knowledge as the main productive force—fully coincides with fixed capital— that is, the 'scientific power' objectified in the system of machinery.

© The Author(s), under exclusive license to Springer Nature
Singapore Pte Ltd. 2021
K. Psarikidou, *Reclaiming the Knowledge Economy*,
https://doi.org/10.1007/978-981-16-6843-2_2

instrument in the hands of labour to a force of production that is increasingly detached from labour or even opposed to it. For him, the introduction of technology in the form of machinery was catalytic in turning 'abstract knowledge' into a powerful mechanism for the separation of labour from the means of production as well as for new divisions within labour. As he stated in *Capital I*, 'The machine ... is a mechanism that, after being in motion, performs with its tools the same operations as the worker formerly did with similar tools' (pp. 492–493). In this case, machines become the expression of a 'conscious application of science' and scientific knowledge comes to coincide with the fixed capital of the system of machinery, which leads to the marginalisation of the variable capital of traditional productive forces such as the labour power, knowledge and experience of workers.[2,3]

A growing number of early-twentieth-century economists also came to underline the centrality of knowledge in capitalist valorisation processes. Coming from a very different standpoint, Friedrich Hayek (1945) was one of the first to identify the incomplete and unorganised nature of knowledge as a crucial source of the problems within the rational economic order. For him, the main problems of designing an efficient economic system should not be attributed to the allocation of available resources, but to the 'fuller use of' a specific resource, 'the existing knowledge', which appears to be 'initially dispersed among many different individuals' (p. 521). At about the same time, another economist and political scientist identified knowledge production and innovation as critical factors for economic change. In *Capitalism, Socialism, and Democracy*, Joseph Schumpeter (1942) introduced the idea of 'creative destruction of capitalism' as the only way of sustaining economic growth and improving quality of life,[4] and knowledge as a key tool for achieving that. More

---

[2] As stated in *Capital I*, 'The man of knowledge and the productive labourer come to be widely divided from each other, and knowledge, instead of remaining the handmaid of labour in the hand of the labourer to increase his productive powers ... has almost everywhere arrayed itself against labour' (pp. 482–483).

[3] In his description of the composition of capital (Capital I, ch. 25), Marx identifies two distinctive forms of capital: constant capital, which is understood as the value of the means of production, and variable capital, which is mainly comprised by living labour power and the total sum of wages, The first is divided into fixed (e.g. long-lasting machinery) and circulating capital (renewable means, e.g. fuels), depending on the durability of the means of production.

[4] According to Schumpeter (1942), the process of 'creative destruction' was regarded as the engine behind economic progress; with the introduction of new ideas and innovation,

recently, Reich (1991) argued that the only true competitive advantage will reside among those who are equipped with the right knowledge for problem-solving. Peter Drucker was another late-twentieth-century economist who underlined the significance of a 'turn' to an immaterial production increasingly dependent on knowledge. As he stated in his *Post-Capitalist Society* (1993), 'the basic economic resource—"the means of production", to use an economist term—is no longer capital, nor natural resources (the economist's "land"), nor "labor". It is and will be knowledge' (p. 8). However, what do all these statements mean? How is knowledge contributing to further capital accumulation within contemporary economies? Has it really transformed traditional production and labour processes?

With the advent of the informatisation processes (Hardt 1999) or what Castells calls the 'information revolution' (Castells 1996), knowledge, information and communication have become central for the reproduction and renovation of capitalism itself through the transition from industry to services and from material to new forms of 'immaterial labour' (Lazzarato 1996). In this context, the traditional industrial factors of production, such as labour, land and capital goods, get supplemented by new immaterial forces, which carry the potential, or at least the promise (as will be discussed below), to further accelerate economic growth. Knowledge and information are not just an external influence on production—as in the traditional production functions that focused on land, labour and capital. Neither are they only introduced as a factor of production—or following a Marxist terminology, a 'productive force'—which could make labour processes, such as the ones involved in agriculture, more productive by decreasing socially necessary production time. As a key element for economic growth, knowledge remains fundamental for increasing production of commodities and labour productivity, but is also transformed itself into a marketable unit. While it is still a resource that can facilitate industrial capitalist production without becoming part of the product, at the same time, it is also amenable to being a product itself that can be distributed, monopolised and sold for the purposes of global economic competitiveness and growth. In this way, knowledge also turns into a commodity with

entrepreneurs could be capable of challenging the existing firms and bring economic growth (Schumpeter 1942). In his later contributions, Schumpeter (1942) paid additional attention to the key role of large firms as engines for economic growth by accumulating non-transferable knowledge in specific technological areas and markets.

an attributed price. The is also explicit in various cases of commercial appropriation and monopolisation of knowledge, such as the appropriation of indigenous people's knowledge, the patenting of the human and plant genome, the subordination of university research to corporate interests and profit motives. In most of these cases, knowledge is not just a commodity, but also 'fictitious commodity' (Jessop 2007), that is, an object which pre-existed in a non-commodified form but is now transformed into a commodity.

So, if knowledge becomes a primary source of productivity and profit, what does this mean for the nature of labour and labour processes? As far back as the first stage of industrialisation, a division between 'productive' and 'unproductive labour' (Marx, Des Kapital, ch.IV) had been crucial for understanding the role of labour and its productivity in capitalist production and valorisation processes. A 'productive labour' was expected to not only meet the needs for reproduction by constantly replacing values that it consumed, but to also contribute to the production of surplus value and capital for the owner of the means of production. Material productivity remains key parameter in evaluating current labour processes. However, with the advent of knowledge as a key element in capital accumulation and growth, labour processes have gradually shifted towards the production of immaterial goods—such as services, knowledge and communication. In this context, 'productivity' is attached not only to the production of material commodities for surplus value, but also to the manipulation of symbols and information, which can also gradually lead to the creation of a surplus value through the production of immaterial goods and services for capitalist appropriation. As a result, 'labour productivity' has also become more dependent on the combination of more complex sets of skills and knowledges, than those involved in industrial labour production. This, of course, does not mean that industrial labour disappeared, but that it just lost the hegemonic role it used to have over in the nineteenth- and twentieth-century economies (Hardt and Negri 2004). As Hardt and Negri claim, although it remains dominant in quantitative terms, a new form of labour emerged that became '*hegemonic in qualitative terms*' (2004, p. 9—emphasis on the original), which, while still mixing with material forms of labour, it also involved the creation of immaterial products—such as knowledge, communication, relationships, emotions (ibid.). It was a new form of labour, also based on a more diverse combination of work skills—intellectual skills, manual skills, entrepreneurial skills—that were more suitable for the processes of 'immaterial production' aiming at the manipulation, production and consumption of information and knowledge outside the traditional production processes[5] (Lazzarato 1996). It thus constituted a new form of labour—what Lazzarato calls 'immaterial labour'—which came to involve new labour processes increasingly based on the human intellect.

Hardt and Negri (2004) identify two principal forms of immaterial labour that are usually combined in the depiction of 'immaterial labour' processes. The first mainly refers to the intellectual or linguistic labour that is aimed to produce ideas, symbols, codes, texts, linguistic figures, images. In this case, production processes not only rely on the productive efficiency of labour to act like a machine both inside and outside the factory and produce material commodities for surplus value. They also depend on the efficiency of the labour to act like computers, and to contribute to a productive communication and continual exchange of information and knowledges, which could then lead to the production of immaterial commodities for surplus value (Hardt 1999). As Hardt underlines, capitalism not only intends to increase the productivity of traditional labour forces, but it also comes to transform some previously non-productive human activities into productive labouring processes. And yet immaterial labour is not only restricted to the manipulation of symbols and knowledge and information for the increase of the surplus value of the commodities, but it also comes to create and manipulate human affect and values, producing and incorporating social relations into capitalist relations of production. In this case, immaterial labour takes the second principal form of an 'affective labour' (Hardt and Negri 2004): a form of labour that, by producing or manipulating affect, communication, social relations and cooperation, not only does it transform human contact and interactions into essential parts of capitalist accumulation processes, but it also contributes to the production of new commodities whose value stems from the new, relational and emotional elements added to the production processes[6] (ibid.; Morini 2007). By combining these two forms, as Hardt and Negri underline (2004), immaterial labour should not be understood as a labour that becomes immaterial. As they explain, "the labour involved in the immaterial

---

[5] For Lazzarato (1996), immaterial labour becomes a mutation of the 'living labour' which operates and affects the society at large and establishes a new social relation, in which not only commodities, but also capital relations are produced in subjective and ideological terms.

[6] Hardt and Negri (2004) discuss the work of flight attendants, health care workers and journalists as fitting into the affective labour category.

production ... remains material—it involves our bodies as all labour does. What is immaterial is *its product*". It is a labour that creates not only material goods, but also relationships and ultimately social life itself' (p. 109—emphasis in original). In doing so, it also contributes to the transformation of the social life itself, by transforming the working conditions but also by blurring the boundaries between work time and leisure time, that is, the economic and the social (Hardt and Negri 2004).

So, how have these complex socio-economic transformations been articulated in more recent economic developments? Is the 'knowledge economy' part of such a vision towards an increasingly dematerialised economy? What do we understand as 'knowledge' in the KBE and what is the importance of knowledge in transforming processes of commodification and labour, as well as visions of growth and productivity within it?

Knowledge Economy: Framing Knowledge and Labour in Contemporary Economies

The concept of the 'knowledge economy' has been coined to depict the centrality of knowledge, information and culture in the pursuit of future economic growth. Many OECD and EC countries urged to develop policies committed to transforming the knowledge-intensive business sector into the most rapidly growing sector in the world (OECD 1996) and the European economy into 'the most competitive and dynamic knowledge-based economy in the world capable of sustainable economic growth with more and better jobs and greater social cohesion (European Commission 2000). However, what is new with the 'knowledge economy'? How is knowledge and labour being understood and framed within the context of a 'knowledge-based economy'?

Elements of a primitive knowledge economy could be identified in the Paleolithic era, when well-informed bodies of knowledge with respect to animal behaviours, pyrotechnology, mining, symbolic communication, the aerodynamic properties of weapons, cosmology and even medicine were crucial components of economic activities (Smith 2000). A similar claim can be made for the economic activities of modern tribal peoples' lives, as well as those involved in the production processes of the industrial revolution. However, according to Godin (2006), the first official conceptualisation of the 'knowledge economy' can be traced back to the 1960s and the rising significance of New Data. In the 1990s, the concept re-emerged to depict the contemporary changes in the economy, primarily led by the application of new Information and Communication Technologies (ICT). Since then, in response to the rising competition of the low-wage industrial economies of the South, the knowledge-based economy provided the panacea for the economies of the Global North—for example, already occupying over 40% of the European workforce in 2005 (Eurostat statistics in Brinkley and Lee 2007).

According to the Work Foundation, 'Knowledge Economy' covers a wide variety of technology and knowledge-based industries, mainly reflecting R&D intensities, high ICT usage and the deployment of large numbers of graduates and professional and associate professional workers. However, all these activities are not coming together for a common definition of the knowledge economy. Based on various EU, OECD and UN statistics for the knowledge economy, Brinkley and Lee (2007) and Walby (2007) have provided a categorisation of different parts of economy that a 'knowledge economy' can encompass. For example, the knowledge economy can comprise the specific industrial economic sectors mainly associated with highly educated workers dealing with Information and Communication Technologies, as well as a range of other high-tech developments, such as this of biotechnology. It has also been associated with high-tech manufacturing, including the manufacture of relevant machineries, as well as the use of ICT in industries, such as telecommunications, software, publishing and data processing. Second, the knowledge economy is associated with a range of knowledge-based industries with an interest in the information sector, more narrowly defined than ICT. This understanding seems to narrow down the gap between manufacturing and services, since it mainly focuses on content industries, such as publishing and media, as well as ICT services such as telecommunications, software publishing and data processing. Third, the knowledge economy is the knowledge-intensive service economy that includes not only high tech services, such as telecommunication and research, but also knowledge-intensive market services such as financial services as well as the education, health and recreational sectors.[7] Despite the inherent ambiguity depicted in the different conceptualisations and the ultimate use of the term, they all seem to be part of a specific rhetoric for identifying new development

---

[7] For a more detailed categorisation of the knowledge-intensive industries and their overall contribution to the knowledge economy and unemployment rates in the EU, see Figure 1 in the Appendix I.

and commodification pathways, which can be 'based on the production, distribution and use of knowledge and information' (OECD 1996, p. 3; Foray and Lundvall 1996).

However, how is knowledge understood within the context of the newly emerging 'knowledge economy'? How does the idea of the knowledge economy make a difference to the traditional understanding of knowledge as a productive force for industrial economies? Following the dominant capitalism-driven economising logic, in the 'knowledge economy', knowledge appears to be primarily associated with the techno-scientific knowledge as a key driver for future productivity and growth. What also becomes important in the KBE context is also the focus on the actual productive potential of knowledge itself. Therefore, knowledge is not only a means for perpetuating the separation of labour from the means of production and thus achieving higher labour productivity through a decrease in the socially necessary labour time. Within the context of the KBE, knowledge is also envisioned to turn into a valuable commodity, which can be produced, distributed and sold for profit. By taking the form of information, knowledge becomes increasingly 'embrained' in highly educated workers and 'encoded' as part of technology-driven science, research and innovation processes. As Gibbons et al. identified (1994), while science and scientific knowledge used to be distinguished from knowledge generated by more applied and commercial research, the boundary between basic and applied research, or else between science and technology becomes less sharp, since scientific and technological knowledge become products of the same Research and Development framework. In this context, knowledge takes the specific form of *exploitable* knowledge, which, in its codified form, is suitable to meet the needs and demands of the 'Industry' for productivity and growth (Abramowitz and David 1996, p. 35).

The 'knowledge economy' has also brought a significant transformation in labour processes, with 'knowledge workers' constituting the new form of 'productive labour' fostering capital accumulation through commodification of knowledge. Defined as workers equipped 'with a considerable theoretical knowledge gained through formal education'[8] (Nisikawa and Tanaka 2007, p. 215; see also Burton-Jones 1999; Drucker 2002),

---

[8] The definition of 'knowledge workers' varies considerably according to author. However, according to the 2007 Work Foundation Report (Rudiger and McVerry 2007), only the 37% of Europe's workers are 'knowledge workers'—defined as those involved in the top three

the 'knowledge workers' have been expected to transform the previously considered 'unproductive' service labour into a productive force, capable of producing surplus value through the manipulation, communication and commercialisation of immaterial goods and services. They are expected to reduce former economic dependence on spatio-materially bound capital through the utilisation of knowledge as new means for wealth creation. They are the new *immaterial* form of highly skilled labour that is increasingly dependent on the production of 'knowledge as commodity' by highly skilled workers. This inevitably results in new divisions of labour between the 'knowledge workers' and 'non-knowledge workers', between 'skilled' and 'unskilled' workers, or else between the immaterial labour, whose value stems from their education, expert thinking and complex communication skills, and the 'material' or 'manual labour', usually displaced in the Global South to be met at a lower cost.

Thus, in the context of knowledge economy, both knowledge and labour get re-defined. Following a Schumpeterian approach (1942), we could claim that this re-configuration of knowledge and labour come to construct the basic elements of a new form of capitalism, one which is animated through the 'creative destruction' of the components of a prior capitalist economic order. However, is the 'knowledge economy' a real economy or is it just a policy imaginary with its own 'prescription for reality' (Ponte 2009)? In many ways, the 'knowledge economy' constitutes another prescriptive rhetoric for future economic growth. Following Jessop (2005), we could approach it as a new master economic narrative that can be easily adopted in many accumulation strategies, state objects and hegemonic visions for the construction of a post-Fordist accumulation regime.[9] On the one hand, knowledge—through its commodification and its transformation from a collective resource into intellectual property,

---

occupational categories (managers, professionals and associate professionals, such as nurses and computer technicians) or graduates. For more details, see Figure 2 in Appendix I.

[9] As opposed to the term *Fordism*—that mainly describes an economy characterised by the separation of ownership and control in large corporations, industrial mass production of standardised goods and stable long-term employment conditions of factory workers able to purchase the products they make and thus increase demand and productivity—*post-Fordism* mainly refers to one characterised by a decline of the old manufacturing base and a shift towards the use of the new 'information and communication technologies' (ICT), flexible specialisation of production in accordance with demands and choices, lifestyle, tastes and culture of different groups of consumers, and the establishment of more flexible, mobile and precarious labour and work organisation conditions.

from a public good to an economic good for profit maximisation—can be seen as the means for a new economic phase that could transform not only social and economic life, but biological life itself. On the other hand, the knowledge economy has been claimed to be simply a buzzword (Godin 2006) for describing a new phase of capitalism, where knowledge is conflated with information, scientific knowledge is conflated with technology and technological advances get conflated with societal change. It has been described as a prescriptive narrative, which mainly directs policy-makers to the adoption and construction of certain policies, institutional changes and research agendas primarily based on the centrality of science and technology for the production of innovation and growth (Ponte 2009; Birch et al. 2010). However, where does the truth lie? In the next section, the case of the knowledge-based bio-economy will be central for initiating a debate over the performative potential of the KBE as a prescriptive narrative shaping reality in its own image.

The Knowledge-based Bioeconomy: Capitalising Knowledge and Life

The concept of the 'knowledge-based bio-economy' is key for understanding the more specific nature and ways in which an agro-food economy can constitute a knowledge economy. The idea of the 'knowledge-based bio-economy' emerged to describe the economic potential opened up around the production of new knowledge around life and other biological processes, including agricultural processes. The term 'bioeconomy' can be traced back to the late 1990s, when biotechnology was promoted as a symbol of a larger 'bioeconomy' aiming at competing with the dual threats of biological vulnerability and economic competition. As underlined in one of the Commission reports, 'the bioeconomy includes all industries and economic sectors that produce, manage and otherwise exploit biological resources ... KBBE will play an important role in a global economy where knowledge is the best way to increase productivity and competitiveness and improve our quality of life, while protecting our environment and the social model' (DG Research 2007). According to the European Commission, it is the 'sustainable, eco-efficient transformation of renewable biological resources into health, food, energy and other industrial products' (DG Research 2006). However, what is new about the bio-economy? Could we not claim that past economies were also knowledge economies based on the knowledge about biological processes and the use of natural resources?

As claimed in one of the DG Research reports (2005), the bioeconomy should not be considered as something new: *'the bioeconomy is one of the oldest economic sectors known to humanity...we have always depended on nature's bounty. In fact, human civilisation is firmly rooted in agriculture. Without the invention of farming, we would not have had the necessary basis for civilisation to bloom'* (p. 2—italics in original). Such observations can also get us back to Marx, who, in his *Economic and Philosophical Manuscripts*, underlined humans' dependence on nature—also described as the process of 'metabolic rift'. For him and Engels, nature constitutes man's inorganic body; 'man lives on nature...nature is his body, with which he remains in a continuous interchange if he is not to die' (Marx and Engels, Collected Works 1987, p. 276, in Benton 1989). In a way, all economic activities have been based on human's constant interaction with nature, the use of nature for the purposes of human flourishing and the enhancement of people's knowledge through an engagement with nature. With the advent of capitalism, nature still constituted both a barrier and an opportunity for growth. As Marx observed, overcoming the biological barriers and obstacles to capitalist expansion and penetration has been one of the key challenges to pursuing capital accumulation. And, in this context, as they identified, agriculture was a significant recalcitrant sector that capitalism needed to address for reinforcing those relations of production that could enable the separation of the worker from the means of production, also through the incorporation of other, off-farm practices into agriculture.[10]

So, what is new within the KBBE? As claimed in the same DG Research report, 'life sciences and biotechnology are transforming it [the bioeconomy] into one of the newest [economic sectors]'...'bioeconomy should not be written off as some outdated notion ... it is also leading change in the twenty-first century and is at the vanguard of the emerging knowledge-based economy'. According to the EU Science and Research Commissioner Janez Potocnik, *'As citizens of planet earth, it is not surprising that we turn*

---

[10] The central role of agriculture in the generation of capitalism has been underlined in *Grundrisse* (Marx 1973), when Marx describes the consequences of a relative subsumption of agriculture within capitalism: 'agriculture no longer finds the natural conditions of its own production within itself, naturally, arisen, spontaneous, and ready to hand, but these exist as an industry separate from it ... and this transfer of the conditions of production outside itself, into a general context ... is the very tendency of capital'.

to Mother Earth—to life itself—to help our economies to develop in a way which should not just enhance our quality of life, but also maintain it for future generations' (ibid., p. 2—italics added). In a way, we could claim that the backward-looking way of describing this economic activity seems to provide a legitimate ground for a forward-looking vision of future economic productivity, reflected in the concept of the bio-economy. The KBBE appears to fit in with past discourses of human economic dependence on nature and life, but it also introduces a step further, through which knowledge enters into the economic calculus of productivity, and science and innovation become means for the materialisation of future economic competitiveness, productivity and growth.

Biotechnology appeared as a key instrument for achieving this bio-economic vision (Rose 2008). This is also evident in most reports on the 'bio-economy', which highlights the centrality of biotechnology as the set of revolutionary techniques that, based on its ability to transform life, can make a significant contribution to the economy. In this context, science, research and innovation are configured as the driving forces for wealth creation, and biotechnology is perceived as the most appropriate type of knowledge which would provide a competitive advantage by 'harness[ing] biological processes for practical implications' (OECD 2003). As claimed in the OECD's Bioeconomy Project, the bioeconomy can be made possible only by the 'recent surge in the scientific knowledge and technical competences … biotechnology and the biosciences more generally have the potential to generate significant economic, social, health and environmental benefits … [through] the unrivalled access to information about biological processes and the ability to process this quickly and link knowledge to outcomes in ways that have previously not been possible' (OECD 2006, pp. 3–4). However, is biotechnology new? How does it relate to activities of the past? How is it different from other forms of knowledge?

For Kloppenburg, 'biotechnology' encompasses human activities of considerable antiquity: 'the manipulation of microorganisms or even more importantly the breeding of plants and animals could be considered as old forms of biotechnology' (1988). However, with the advent of new Genetics, as Thorpe comments (2011), plant breeders started to 'make natural history themselves', working with the natural limits of sexual compatibility. In this context, the 'decontextualisation, reification and commodification of the productive and reproductive capacities of living things' (Thorpe 2011 in Birch et al. 2010) is transferred to a different level, where 'humanity acquires a qualitative superiority over conventional methods of genetic manipulation and the alteration of the living organisms' (Kloppenburg 1988). This carries the potential to not only exploit the natural resources for the production of new commodities, not only transform natural resources into material commodities for capitalist accumulation, but also transform natural resources into a new form of non-human productive labour that can complement but also fill in the potential deficits of human labour productivity. Within the bioeconomic framework, biotechnology appears to hold the promise to create and reproduce the material conditions of our existence through the (re)making of nature into a productive force (Brennan 2000), which can 'work harder, faster and better' (Boyd et al. 2001, pp. 563–564) for the pursuit of surplus value.

In this context, on the one hand, 'life itself' is not only genetically transformed into something new, but also gets perceived as a piece of information, as DNA molecules in labs, which can be packaged, turned into commodity and sold in a database form. In other words, the commodification of 'life itself' is also enabled through the codification of its material reality, through which the meaning and grammar of 'life-as-information' becomes easily interpreted in terms of a calculable market unit with a commercial value for contemporary market economies (Thacker 2005; Sunder Rajan 2006). In this context, both knowledge and nature are seen as resources for the reproduction of capitalism. They both become translatable into economic value. Within KBBE, surplus value not only stems from the decrease of socially necessary production time and the increasing productivity of labour. On the one hand, nature is transformed into a 'resource' through the production of new knowledge, but it also facilitates the production of new knowledge, which can then turn into commodity. Nature and biological processes are configured as new labour forces, contributing to the production of surplus value for capitalist accumulation. In this way, the KBBE contributes to an increased surplus value through the manipulation of nature as the new productive labour that would increase independence from human labour and its limits to productivity (Brennan 2000). By capturing 'life as information', it enables the regeneration of capitalism through its capacity to 'produce productivity' (Cooper 2008). By making nature more productive, not only does it overcome its dependence on the limited socially necessary labour time, but also succeeds in also reducing the 'naturally necessary labour time' for capital accumulation (Brennan 2000).

Along these lines, the KBBE appears to construct a techno-knowledge fix vision, which, as Birch et al. claim (2010), primarily respond to the neoliberal agendas for the creation of a 'more sustainable' capital or, based on the co-production of the life sciences and capitalism (Sunder Rajan 2006). In this case, knowledge appears to be reduced to techno-scientific knowledge. Innovation comes to be conflated with technological innovation; in many cases, techno-scientific innovation also gets conflated with biotechnological innovation. Under these conditions, nature turns into a 'resource' configured around specific narratives and knowledges, promises and visions that result in justifying certain policy and research agendas, while marginalising others. However, what are these promises and visions, and what are their implications for the materialisation of the KBBE as a master economic narrative? The case study of the KBBE in the agro-food sector offers some fertile ground for exploring the KBBE narrative as one of reality or hype.

The KBBE in Practice: The Case of the Agro-Food Sector

The agricultural sector constitutes an interesting sphere for observing such narrow techno-scientific understandings of knowledge and innovation within the context of the KBBE. Agricultural bio-technology appears to have a prominent role in this vision. As evident in OECD'S 2006 report as well as the 2002 Life Sciences Strategy, biotechnologies are configured as the technological solution that would address the genetic deficiencies of nature, as well as other socio-environmental limitations of prior forms of green revolution. Based primarily on the exploitation of the commercial potential of nature and its genetic information, it is configured as the model that would help address health, nutrition and environmental matters, while promising to bring both wealth creation and quality of life. As said in the EC Life Sciences Strategy:

> In the agro-food area, biotechnology has the potential to deliver improved food quality and environmental benefits through agronomically improved crops ... Food and feed quality may be linked to disease prevention and reduced health risks ... Life sciences and biotechnology are likely to be one of the important tools in fighting hunger and malnutrition and feeding an increasing human population on the currently cultivated land area, with reduced environmental impact. (2002, p. 11)

However, as also stated in the Genewatch UK report *Bioscience for Life* (Wallace 2010), in most of these policy documents, significant future benefits are described and promoted rather than actually evaluated. They adopt a strong 'vision-led' approach, which mainly aims at influencing political decisions and policy-making in the agrifood domain. Most of the statements are based on claims about high-yielding varieties which do not yet exist, but are promised to be developed in the future as a result of public investment in research and development, primarily for genetically engineered crops. In this context, the bioeconomy appears to become the enabler for the generation of commercial value in the present in the promise of making a certain future possible. For example, according to the Biopolis report, between 2002 and 2005, the total public funding of biotechnology in the EU was estimated to 13.1 billion euros (Wallace 2010). As for agriculture, the Prime Minister's Unit reported that estimates of global GM crop R&D (Research & Development) were about $4.4 billion a year and the European Commission estimated that the EU spent 80 million euros annually on research in plant biotechnology. More specifically, along with the EC FP7 'Plants for the Future', the agricultural biotech industry required another 45 billion euros for R&D between 2010 and 2020.

Wallace (2010) has commented on the ambiguity of the promissory vision of the agbiotech sector. For her, the KBBE agenda succeeded in transferring commodification processes from the agricultural field to the scientific laboratory. In this case, capitalism appears to re-invent itself by succeeding in overcoming potential prior obstacles to capitalist accumulation. Profit-maximisation is not just the outcome of commodification of the fruits of agricultural production, but also of the new scientific knowledge produced around agricultural production; a scientific knowledge that promises to re-invent nature and thus overcome the prior natural limits of productivity within agricultural production. In this context, commodification not only refers to commodification of natural resources, but also to the commodification of knowledge around the agro-food processes. It refers to the manipulation of life and biological processes for the generation of surplus value associated with the commoditisation of both plants and knowledge as resources. In this context, science also appears to serve the purposes of the agro-biotech industry. Research priorities shift from farm to laboratories, and laboratory knowledge appears to be prioritised at the expense of other agricultural knowledge systems that get marginalised, redefined and lost (see also Wallace 2010). In this context, wealth creation also gets narrowly defined, mainly around the profits generated through the production and commodification of laboratory knowledge.

The KBBE becomes part of a vision-led approach that encourages investment in particular research priorities that are promised to make a certain 'productivist' agricultural future possible—thus, also resulting in shaping agrifood research and knowledge production in certain ways that could serve this vision. Surplus value seems to increasingly stem from the commodification of immaterial goods and services developed around scientific research processes. Wealth creation becomes configured around laboratory-knowledge creation. Science and knowledge are primarily configured around the interests of the biotech industry and its promised techno-fixes to former agricultural deficits. Knowledge becomes highly technical and captured in a form, which could enhance its commercial and patentability potential. Their intangible abstractions for future productivity and profit seem to work as tools for the reproduction of capitalism, whose power, as Marx underlined (1976 [1867]), is hidden in the magical nature and capacity of these commodities to elude capture in purely materialistic terms.

This Schumpeterian vision of growth also comes with inevitable changes at a policy level. On the one hand, policies appear to divert support and investment towards these new promissory spaces for knowledge-driven economic growth, thus diverting their focus from those directly related agricultural policies. As early as 2007, the 'En-route to the Knowledge-based Bioeconomy' report provides evidence of this change: 'agricultural policy is not in line with the goals of sustainable production and healthy food … The heavy dependence on agricultural subsidies is expected to decline … shifting subsidies from the Common Agricultural Policy (CAP) to support science and research programmes, e.g. on sustainable agriculture and new 'industrial crops' is suggested since the integrated production of food, fibre and energy could become profitable in their own right' (1942, pp. 4, 15). Thus, the knowledge-based bioeconomy appeared to provide the space for the justification of particular policies which could engender certain conditions in present that could make certain agricultural techno-scientific futures possible. However, as a result of this, other policy frameworks get marginalised or constrained, whereas research funding is shifted from the farm and the enhancement of the farming knowledges and skills to the laboratory and the enhancement of techno-scientific knowledge, primarily for the economic benefit of the agro-food and biotech industries.

In this context, scientific knowledge and innovation become key in the configuration of certain agro-food imaginaries. Imagined futures are not so much focusing on the agro-food sector itself, but on the generation and accumulation of profit from and for the laboratory-based knowledge industry. In this way, following Sunder Rajan (2006), we could claim that both genomics and biotechnology become key components of a promissory 'game' played in the future in order to generate the present that will enable this future. And, within this 'game', the KBBE policies come to play a significant role for making a certain political bio-economic future possible.

The KBBE Narrative: Between Truth, Promise and Hype

According to Rajan (2006), the words 'life', capital', 'fact', 'exchange' and 'value' are integral part of a lexicon describing the conditions of a new bio-based economic order. This is also evident in the OECD report 'The Bioeconomy to 2030: Designing a Policy Agenda' (2009), which identified capital-intensive biotechnological innovation as the solution to the global socio-economic and environmental challenges societies face (e.g. rapid increase of the population, a rising demand for and depletion of essential natural resources, global warming and climate change, deforestation). In this context, natural resources become part of a certain image constructed around narratives of a deficient vulnerable present (Cooper 2007), whose remedies lie in the construction of a Knowledge Biosociety (Gottweis 1998). The KBBE constitutes such an alternative agenda that promises to avoid, or even provide a remedy towards, past problems attributed to the genetic deficiencies of human and non-human nature. It aims to overcome the limits of socially and naturally necessary labour time through the production of new commodities that can modify life itself (Brennan 2000). In doing so, it succeeds not only in producing surplus value, but also in overcoming the ecological contradiction of previous capitalist economic forms—primarily through a partial dematerialisation of the economy and the introduction of new forms of immaterial labour and its commodities.

Therefore, the KBBE agenda is promised to bring about social and economic change through the production of biovalue (Waldby 2002), a value also linked to its promised capacity to correct failures of vitality and other genetic deficiencies prone to biotechnological treatment (Birch 2007, p. 94; cf. Rose 2008, p. 21). It thus contributes to the generation of new kinds of capital flows and commodities, whose value derives from future economic returns for the bio-knowledge inventors. In the hope of creating surplus value, the KBBE results in absorbing the use value of various biological resources, and producing new use and exchange values related to their treatment as commodities to be bought and sold for profit. However, it also enhances profit through the commodification and privatisation of the laboratory knowledges around certain genetic characteristics, which can then also provide higher-value inputs and outputs (Levidow 2008).

In this case, in the bioeconomy, 'appearing to succeed becomes more important than succeeding' (see Ponte 2009, p. 484), as, most of the times, returns appear to be made in the speculative phase of scientific research, regardless of whether or not any usable commodities will result. However, even if a promise comes to fulfilment, this is not so much the outcome of a well-founded prediction, but the result of the performative effects of a particular, partial representation of a 'situation' and what comes to either constitute or be needed as part of it. Thus, the KB(B)E constitutes another 'self-fulfilling prophecy' (Merton 1948), a specific descriptive narrative of a situation which succeeds in becoming part of this situation and further affecting subsequent developments. In doing so, it also succeeds in creating a partial or even false definition of a situation, generating specific behaviours that may result in a distorted conception of reality becoming true. Thus, by adopting a specific language that would bring about its prediction about future economic development, the bioeconomy succeeds in creating a specific conception of truth, thereby signalling a new phase of capitalism necessitated by biological morbidity and genetic remedies. As Sunder Rajan claims, it introduces 'a new face and a new phase of capitalism'—a continuation of, an evolution of, a subset of, and a form of capitalism—which comes from the creative destruction of former economic sectors that, within the context of new global industrial economy, appear weak to maintain their competitive advantage. It could be seen as 'a case study' of the co-production of techno-scientific and economic regimes, where life sciences, emergent biotechnologies and the marketplace work together for the generation of new commercial values given in the form of biological material as well as the production and commodification of knowledge and information about it.

In this context, the KBBE is configured as an *economic imaginary*[11] which, as Jessop describes (2005), leads to the consolidation of a certain accumulation regime and regulatory framework that succeed in shaping socio-economic realities in their own image. It is configured as a new dominant economic narrative capable of creating the framework within which techno-scientific visions can be articulated and take place. However, we could also claim that it succeeds in going beyond an imaginary framework with regard to its own existence. Through the articulation of certain promissory visions for the future, it succeeds in making itself true in present. It thus succeeds in not only bringing certain socio-economic realities into life, but also becoming part of this promised socio-economic reality, on the basis of which a credible futuristic vision, or else a 'forward-looking statement' (Rajan 2006), can be promoted and sold.

In this context, the KBBE constitutes an economic imaginary which succeeds in shaping socio-economic realities by promoting a futuristic vision which does not have to be true. It is the form of capitalism that Susan Strange (1986) describes as 'casino capitalism': aiming at the 'creation of the new, new thing' (Lewis 1999) and the generation of commercial value in the present, in an attempt to make a certain future possible. Following Rajan, the KBBE can be understood as a hype, which 'is not about truth and falsity; rather, it is about credibility and incredibility'. It is a rhetoric that brings about specific future techno-scientific imaginaries tied to particular research, industry and policy agendas and their visions

[11] The concept of imaginary—usually associated with the idea of *social imaginary*—has also been analysed by various scholars to describe a guiding tool that mobilises social subjects in a world of uncertainty (Gonzalez-Velez 2002). Although Jessop's conceptualisation of the economic imaginary seems to better fit in the description of the KBE/KBBE, an account of the different conceptualisations of the social imaginary shows the interconnection between terms which stress the role of different means—for example ideas, practices, discourses—through which social subjects organise the social conditions and collective forms of their existence. Cornelius Castoriadis (1987), Michael Maffesoli (1993a, 1993b), Charles Taylor, and Arjun Appadurai (1996) have all contributed to developing a better understanding of the concept and its role in constructing a collective conscience. For Castoriadis, it constitutes a set of imagined means through which social subjects could overcome the conformity to discursive dispositions already made up by the conscience collective. Maffesoli focused on the sense of solidarity that the social imaginary gives to social subjects, whereas Taylor approached it as a common understanding which enables people in society to carry out collective practices that make up their social life. Through all these different approaches, according to Gonzalez-Velez (2002), the social imaginary appears as 'a taken for granted truth, that gives unity and order to social subjects lives and facilitates the continuity of a collective that is fragmented' (p. 351).

for economic growth. What really counts is the ways in which these promissory visions for the future create the conditions of possibility for the existence of the biotech industry in the present. Thus, it does not matter whether the KBBE promissory visions are true or not, as long as they are credible.[12] What really matters is selling the visions of future products, and thus turning those visions into their products, rather than selling the products themselves, since profit does not really stem from those material commodities but from the increasingly dematerialised labour and knowledge commodification processes produced around it.

In this sense, the bioeconomy is a master economic prescriptive narrative, which succeeds in constituting part of socio-economic realities of the present. By introducing some specific future visions, it succeeds in 'shaping [present] reality in its own image' (Ponte 2009, p. 485). Although it has been claimed to constitute both a techno-scientific and a socio-technical imaginary, which enables the constitution and production of new imagined forms of social life and economic order—an imaginary which is not only tied to particular scientific and technological projects, but is also imbued with an implicit understanding of the social world, which it aims to encode, conceptualise and redesign through these projects (see Jasanoff et al. 2008). However, it can also be seen as going beyond its 'imaginary' status and constituting this new economic narrative, which is 'at once descriptive of attainable futures and prescriptive of the kinds of futures that ought to be attained' (Jasanoff et al. 2008). Following Merton (1948), it is configured as a self-fulfilling prophecy—the 'messianic space' which, in the promise of the production of use values in the future, makes itself true and contributes to the production of surplus value in the present through the transformation of nature and knowledge into both productive forces and commodities operating within the broader context of the social relations of production (cf. Sunder Rajan 2006; Birch et al. 2010; Ponte 2009; Cooper 2008).

---

[12] According to Derrida, it is impossible to locate an unfulfilled promise as a lie even as it is always perceived as nontruth: 'It will always be impossible to prove in the strict sense that someone has lied even if one can prove that he or she did not tell the truth' (Derrida 2001, p. 68).

Conclusions

The discussion in this chapter reveals that, both knowledge-based economy (KBE) and the knowledge-based bio-economy (KBBE) become pivotal in re-inventing capitalism through the introduction of a new phase and face of capitalism that would overcome the economic and environmental limitations of previous economic forms.

Despite the historic origins of the term, 'knowledge economy' has recently attracted a lot of attention as part of a vision of an increasingly dematerialised economy that could contribute to the re-production and re-innovation of capitalism. In this context, knowledge is transformed into a new factor of production, as well as a commodity, which in its codified form can easily be packaged, distributed and sold for the purposes of economic growth and profit. New immaterial and affective forms of labour, sometimes also described as 'knowledge workers', also come into play contributing to the incorporation of previous 'unproductive' forms of labour in the social relations of production.

With the advent of the knowledge-based bioeconomy narrative, the generation of surplus value also accelerates through further processes of economic dematerialisation. The KBBE provides a new promissory space for economic growth, based on the production as well as commodification of knowledge for the purposes of genetic manipulation, codification and commodification of nature, life and other biological processes and resources. In this case, capitalism is constrained to not only changing the social relations of production around nature, but also changing nature itself. Nature is configured as the productive force that can increasingly detach the production of surplus value from human labour processes. And, the agricultural sector provides a unique opportunity for the materialisation of such processes.

The KB(B)E becomes a descriptive *and* prescriptive narrative of certain futures that are devoid of the natural limitations of capital accumulation related to human and natural deficiencies of the past. It constructs its own promissory discourse and, based on a promise of a future use value, it aims to generate present commercial value and economic returns attached to the exploitation of knowledge and life as productive forces *and* resources. It can be seen as a rhetoric, which succeeds in supporting certain economic and knowledge sectors, primarily based on the promised potential future benefits they can bring to wider communities and publics. In other words, it constitutes a speculative promise—in the way it can produce good results in investment even if nothing results in the long term. And, in doing so, in many cases, it can also turn into a self-fulfilling prophecy, primarily based on a distorted representation of a situation that succeeds in creating a specific conception of truth, which, in some cases, it can even bring into being. In this way, it also constitutes an imaginary whose credible futuristic vision makes itself true by bringing into being certain socio-economic realities, policy and research agendas, while at the same time marginalising other imaginaries and visions of the futures that could promote an alternative research and policy agendas.

In this way, the KBE narrative becomes subordinate to a specific policy and research context, which appears to serve this dominant knowledge-economic narrative constructed around growth. However, could the KBE make sense in a different political context? Could we identify different ways an economy can be called a knowledge economy? And, what would this mean for the articulation of different policy and research agendas, which now seem to be marginalised behind the dominant understanding of the KBE master narrative? As outlined above, a historical interdependence between different types of economy and knowledge becomes prevalent. In this way, all economies could be claimed to be knowledge economies. With the contemporary framework of the knowledge economy as a messianic space and imaginary for future economic growth, we could claim that the KBE has been dominated by a particular understanding which could only make sense along the lines of a techno-scientific and capitalocentric vision of innovation and growth. However, why should we not try to understand the knowledge economy beyond such narrow capitalocentric and technocentric frameworks of thinking? And, what would have been the implications of that in terms of future research and policy?

As we discussed above, the agro-food sector constitutes a very interesting case for understanding the centrality as well as the character of the knowledge economy as a new phase and face of capitalist post-industrial economic narrative, vision and reality. It has been key for realising the significance of both knowledge and life as both productive forces and commodities for the production of an increasingly dematerialised economy that is capable of also going beyond the environmental deficits of prior economic forms. However, it is also key for realising the rise of a new economy, which is increasingly dependent on promise, and complex commodification processes whose profit is increasingly based on the codification and commodification of knowledge and other fictitious commodities that existed in a non-commodified form. And, within this context, alternative agro-food networks (AAFNs) provide an interesting space for exploring the possibility of alternative knowledge economic frameworks and visions.

In the following chapters, we aim to investigate these particular dimensions and discuss what means to call AAFNs as a KB(B)E and what this would imply for a re-definition of KB(B)E beyond a purely technocratic and capitalocentric understanding. In doing so, certain questions also become key. For example: (a) how is knowledge perceived in AAFNs and how does the knowledge production take place? (b) what are the different knowledge patterns and production processes involved in the AAFN economy? (c) are there any knowledge workers or other forms of immaterial labour involved in their social relations of production? However, before we get into a deeper investigation of the potential KB(B)E dimensions of the AAFNs, we first need to identify the ways and the elements through which we can understand and investigate an agro-food economy as a knowledge economy. In the following chapter, I explore the diverse knowledge forms and knowledge production processes that are important in the configuration of both 'knowledge economies' and 'agro-food economies'. Comparing the commonalities and differences between these two economic sectors is pivotal for not only understanding the agro-food economy as a knowledge economy (which is the aim of this following chapter), but also for shedding light on the key aspects necessary for carrying out a similar investigation in the case of AAFNs (which will be conducted in Chap. 4). Thus, it will provide basis for an investigation that will help us not only understand the AAFNs as a KB(B)E but also reclaim the concept of KB(B)E, and consider further developments that can be opened up around such a possibility.

REFERENCES

Abramowitz, M. and David, P. (1996) Technological Change, Intangible Investments and Growth in the Knowledge-Based Economy: The US Historical Experience. In Foray, D. and Lundvall, B.A. (Eds.) *Employment and Growth in the Knowledge-based Economy*. Paris: OECD: pp. 35–60.

Appadurai, A. (1996) *Modernity at large: Cultural dimensions of globalization*. Minneapolis: University of Minnesota Press.

Benton, T. (1989) Marxism and natural limits: an ecological critique and reconstruction. *New Left Review*, 178:51–86.

Birch, K. (2007) Knowledge, Place and Power: Conceptualising Value Creation in Knowledge-Based Commodity Chains. Working paper. Centre for Public Policy for Regions. University of Glasgow.

Birch, K, Levidow, L. and Papaioannou, T. (2010) Sustainable Capital? The Neoliberalisation of Nature in the European Knowledge-based Bio-economy, *Sustainability* 2: 2898–2918.

Boyd, W, Prudham, S. and Schurman, R. (2001) Industrial dynamics and the problem of nature. *Society and Natural Resources*, 14:555–570.

Brennan, T. (2000) *Exhausting Modernity: Grounds for a new economy*. London and New York: Routledge.

Brinkley, I. and Lee, N. (2007) *The Knowledge economy in Europe. A Report prepared for the 2007 EU Spring Council*. The Work Foundation. Available online at: http://www.theworkfoundation.com/assets/docs/publications/80_Knowledge%20Economy%20EU%20Spring%20Council.pdf.

Burton-Jones, A. (1999) *Knowledge Capitalism. Business, Work and Learning in the New Economy*. Oxford: Oxford University Press.

Castells, M, (1996). *The Rise of the Network Society*. Oxford: Blackwell.

Castoriadis, C. (1987) *The imaginary institution of society*. Translated by Kathleen Blamey. Cambridge, MA: MIT Press.

Cooper, M. (2007) Life, autopoiesis, debt: inventing the bioeconomy, *Distinktion* 14: 25–43.

Cooper, M. (2008) *Life as Surplus: Biotechnology and Capitalism in the Neoliberal Era*. Seattle: University of Washington Press.

Derrida, J. (2001) History of the Lie: Prolegomena. In Rand, R. (ed.) *Futures of Jacques Derrida*. Stanford: Stanford University Press: pp. 65–98.

DG Research (2005) New Perspectives on the Knowledge-Based Bio-Economy: Conference Report (Brussels: DG Research). Available at: http://ec.europa.eu/research/conferences/2005/kbb/report_en.html.

DG Research (2006) Framework Programme 7, Theme 2: Food, Agriculture, Fisheries and Biotechnology (FAFB), 2007 Work Programme, Commission of the European Communities: Brussels, Belgium, 2006.

DG Research (2007) FP7 'Presentation on KBBE' (Brussels: DG Research), Available at: ftp://ftp.cordis.europa.eu/pub/fp7/kbbe/docs/about-kbbe.pdf. Retrieved October 15, 2009.

Drucker, P.F. (1993) *Post-Capitalist Society*. New York: Harper Collins.

Drucker, P.F. (2002) *Managing in the Next Society*. New York: St. Martin's Griffin.

EC (2000) *The Lisbon European Council—An Agenda of Economic and Social Renewal for Europe*. Brussels: The European Commission, DOC/00/7.

EC (2002) Life Sciences and Biotechnology: A Strategy for Europe. COM(2002) 7. Available at: http://ec.europa.eu/biotechnology/pdf/com2002-27_en.pdf.

Foray, D. and Lundvall, B-A. (1996) *Employment and Growth in the Knowledge-Based Economy*. Paris: OECD.

Gibbons, M., C. Limoges, H. Nowotny, S. Schwartzmann, P. Scott and M. Trow (1994) *The new production of knowledge: the dynamics of science and research in contemporary societies.* London: Sage.

Godin, B. (2006) The Knowledge-Based Economy: Conceptual Framework or Buzzword?, *The Journal of Technology Transfer* 31(1): 17–30.

Gonzalez-Velez, M. (2002) Assessing the Conceptual Use of Social imagination in Media Research *Journal of Communication Inquiry*, 26: 349–353.

Gottweis, H. (1998) *Governing Molecules: The Discursive Politics of Genetic Engineering in Europe and the United States.* MIT Press.

Hardt, M. (1999) Affective Labour, *Boundary 2* 26(2): 89–100.

Hardt, M. and Negri, A. (2004) *Multitude: War and Democracy at the Age of Empire.* New York: The Penguin Press.

Hayek, F. (1945) The use of Knowledge in Society, *American Economic Review* 35(4):519–30.

Jasanoff, S., Kim, S-H. and Sperling, St. (2008) Sociotechnical Imaginaries and Science and Technology Policy: A Cross-National Comparison. Project Summary.

Jessop, B. (2005) Cultural Political Economy, the Knowledge-based Economy and the State. In Barry, A. and Slater, D. (eds.) The technological Economy. Routledge. pp. 142–164.

Jessop, B. (2007) Knowledge as a fictitious commodity : insights and limits of a Polanyian perspective. In: Bugra, Ayse and Agartan, Kaan, (eds.) *Reading Karl Polanyi for the twenty-first century: market economy as political project.* Palgrave, Basingstoke.

Kloppenburg, J. (1988) *First The Seed: The Political Economy of Plant Biotechnology, 1492–2000.* Cambridge University Press.

Lazzarato, M. (1996) Immaterial Labor. In Hardt, M. and Virno, P. (eds.) *Radical Thought in Italy: A Potential Politics,* Minneapolis: University of Minnesota: pp. 133–50.

Levidow, L. (2008) European quality agriculture as an alternative bio-economy. In Guido Ruivenkamp, Shuji Hisano and Joost Jongerden (Eds.) *Reconstructing Biotechnologies: Critical Social Analyses.* Wageningen Academic: pp. 185–205.

Lewis, M. (1999) The new new thing: A Silicon Valley Story. New York: W.W.Norton.

Maffesoli, M. (1993a) Introduction, *Current Sociology* 41(2): 60–67.

Maffesoli, M. (1993b) The imaginary and the sacred in Durkheim's sociology, *Current Sociology* 41 (2): 1–5.

Marx, K. (1973) *Grundisse.* New York, NY: Vintage Books.

Marx, K. (1976) Capital: A Critique of Political Economy, Vol. 1. Harmondsworth: Penguin.

Marx, K. and Engels, F. (1987) *Collected Works,* vol. 3. London.

Merton, R.K. (1948) The Self-Fulfilling Prophecy, *The Antioch Review* 8(2): 193–210.

Morini, C. (2007) The Feminisation of Labour in Cognitive Capitalism, *Feminist Review* 87: 40–59.
Nisikawa, M. and Tanaka, K. (2007) Are Care-Workers Knowledge-Workers? In Walby, S., Gottfried, H., Gottschall, K. and Osawa, M. (eds.) *Gendering the Knowledge Economy: Comparative Perspectives.* Palgrave Macmillan: pp. 207–227.
OECD (1996) *The Knowledge-based Economy.* Paris: OECD.
OECD (2003), *Harnessing Markets for Biodiversity: Towards Conservation and Sustainable Use.* Paris.
OECD (2006) *The Bioeconomy to 2030: Designing a Policy Agenda.* Scoping Paper. Informal Experts' Meeting. OECD Paris, 6 March 2006.
Ponte, S. (2009) From Fishery to Fork: Food Safety and Sustainability in the Knowledge-Based Bioeconomy, *Science as Culture* 18(4):483–495.
Reich, R. (1991) *The Work of Nations: Preparing Ourselves for the 21st Century Capitalism.* London: Simon and Schuster.
Rose, N. (2008) The Value of Life: Somatic Ethics and the Spirit of Biocapital, *Daedalus* 137(1): 36–48.
Rudiger, K. and McVerry, A. (2007) *Exploiting Europe's Knowledge Potential: 'Good Work' or 'Could do better'. A Report prepared for the Knowledge Economy Programme November 2007.* The Work Foundation. Available online at http://www.agoratalentia.es/documentos/trabajadoresdeleconomiadelconocimiento.pdf.
Schumpeter, J. (1942) *Capitalism, Socialism and Democracy.* New York: Harper.
Smith, K. (2000) What is the 'knowledge economy'? Knowledge-intensive industries and distributed knowledge bases. Paper presented to DRUID Summer Conference on The Learning Economy—Firms, Regions and Nation Specific Institutions. June 15–17, 2000.
Strange, S. (1986) *Casino Capitalism.* Oxford: Blackwell.
Sunder Rajan, K. (2006) *Biocapital: The Constitution of Postgenomic Life.* Durham and London: Duke University Press.
Thacker, E. (2005) *The Global Genome: Biotechnology, Politics and Culture.* Cambridge, MA: MIT Press.
Thorpe, C. (2011) Artificial life on a dead planet. *Science as Culture*, in preparation.
Walby, S. (2007) Introduction: Theorising the Gendering of the Knowledge-based Economy. In Walby et al (eds.) *Gendering the Knowledge Economy: Comparative Perspectives.* Palgrave MacMillan: pp. 3–50.
Waldby, C. (2002) Stem Cell, Tissue Cultures and the Production of Biovalue, *Health* 6(3): 305–323.
Wallace, H. (2010) Bioscience for Life? Who Decides what research is done in health and agriculture? Genewatch UK Report. March 2010.

CHAPTER 3

Understanding the Ago-Food Economy as a Knowledge Economy

This chapter investigates the ways in which an agro-food economy is a knowledge economy. Within mainstream economics, scientific and other forms of explicit knowledge are the objective, universal forms of knowledge that are superior to other forms of knowledge. However, to what extent is this true? This chapter aims to unpack the more diverse types of knowledge and more complex knowledge production processes that are important in the making of contemporary economies. It specifically looks into the knowledge forms and knowledge production processes involved in both the knowledge economy and the agro-food economy, and identifies the similarities and differences, as well as the final common threads, which could encourage us to approach an agro-food economy as a knowledge economy. Through such an analysis, this chapter also provides us with some useful methodological tools and units of analyses—for example, types and patterns of knowledge and knowledge production processes— that can help us further investigate AAFNs as an alternative Knowledge Economy in the chapter to follow.

Science, Explicit and Tacit Forms of Knowledge: A Clash or Co-production?

What we call modern science is itself a historical product of a continuous struggle not only to define science in a particular way, but also to exclude other ways of producing knowledge from that definition. (Kloppenburg 1991, p. 24)

The above quote encourages us to understand that what we understand and describe as 'science' is not a unified or single category of knowledge. However, it is also important in realising the more complex power relations and hierarchies that become important in not only framing 'science' in certain ways, but also marginalising other types of knowledges that can be equally important in our understanding and the making of 'science'. As has already been observed in the previous chapter, science and its technological applications have been key for the formulation of a new increasingly dematerialised economic order and a 'knowledge society' driven by innovation. This is what Tovey (2008) came to describe as a 'vision of a future dystopia' (p. 187), where explicit scientific forms of knowledge become dominant in the making of such 'knowledge societies', whereas tacit, situated and embodied, forms of knowledge seem to lose their significance in shaping 'the economic'. However, what is the role of explicit and tacit forms of knowledge in contemporary societies? What is the relationship between them? Could we really argue for a dichotomy between explicit and tacit forms of knowledge?

From a reductionist perspective, knowledge is primarily associated with 'scientific forms of knowledge', understood as a set of static, objective and universal descriptions of the natural processes with a reference to facts, theories and laws. It increasingly comes to be understood as 'justified true belief', perpetuating a separation between the knower and the known, between the knowledge and the subject. It is equated with the accumulation of different pieces of information (Amin and Cohedent 2004), which can be transformed into an object with measurable characteristics that can easily be owned, transferred, stored and valorised in the form of a resource (Smith 2000). This type of 'knowledge as information' takes the form of technically and commercially organised 'knowledge-data' (Nowotny et al. 2001), which not only comes to replace cultural values, but also manipulate nature through the conceptualisation of a de-contextualised, simplified, neutralised, de-socialised and de-politicised nature, which is primarily conceived as a series of problems that need solving. In this way, as Haraway comments (1991), knowledge not only seems to hold a mirror up to reality, but also constructs reality, while allowing prediction and control. It succeeds in bringing new phenomena into existence through the production of standardised material applications and generalisable information.

Within this rising context of scientific objectivity, tacit knowledge—usually described as taking the forms of local, traditional or lay knowledge, 'non-expert' or 'non-professional' knowledge—comes to be silenced, or at best understood as 'information', 'item' or 'items' of quantifiable and substantive knowledge offered for commodity exchange. Following a Kuhnian account of science as a normalising mechanism (1962), knowledge about the natural world is equated with accuracy of prediction and solution of different problems. Science, thus, identifies those important problems and demonstrates how certain of these problems can be successfully resolved through an accumulation of knowledge and the establishment of an 'objective truth' about the natural world (Dreyfus and Rabinow 1982). In this context, tacit knowledge seems to maintain a supplementary role that can contribute to the establishment of these value-free, objective descriptions of natural phenomena. However, to what extent is this the case? Is tacit knowledge disappearing or getting absorbed by scientific knowledge? What is their role in the production of new knowledge?

Michel Polanyi (1967) argued against the dichotomisation between explicit and tacit forms of knowledge. His statement that 'most ... knowledge cannot be out into words' (ibid., p. 4) and that any knowledge is tacit or rooted in tacit knowledge, indicates the complex interconnections between tacit and explicit forms of knowledge. At the same time, he aimed to encourage a re-thinking of the role of tacit knowledge in knowledge production processes, since, primarily situating 'truth discovery' in the pre-logical phases or tacit knowing. As he claimed, knowledge production presupposes a personal responsibility for the pursuit of a hidden truth, or else, the capacity of an individual to relate evidence to a pre-existing non-investigated hidden reality that needs to be discovered[1] (Polanyi 1967).

Therefore, Polanyi aimed to challenge the traditional prioritisation of explicit knowledge and provoke a different way of thinking that would embed the production of new knowledge in tacit knowledge. As he said, 'since a problem can be known only tacitly, our knowledge of it can be recognised as valid only by accepting the validity of tacit knowing' (1967, p. 87). And, he added, 'to acknowledge tacit thought as an indispensable element of all knowing and as the ultimate mental power by which all explicit power is endowed with meaning, is to deny the possibility that each succeeding generation, let alone each member of it, should critically test all the teachings in which it is brought up' (Polanyi 1967, p. 60). In this context, tacit knowing remains an integral part of innovation—since

---

[1] Polanyi underlines the multidimensional nature of reality, by saying that 'what we perceive is an aspect of reality, and aspects of reality are clues to boundless undisclosed and perhaps yet unthinkable experiences' (Polanyi 1967, p. 68).

the latter can only be achieved on the grounds of an unrevealed reality observed by all the knowing actors—whereas explicit knowledge could only be seen as the outcome of the process of knowing, within which the obtained knowledge becomes transmittable in a formal, systematic language.

Polanyi's approach encourages us to re-think the role of tacit knowledge in knowledge-production processes, as well as the boundaries and interrelationship between explicit and tacit forms of knowledge. Codified knowledge—recognised as an explicit, formal or systematic form of knowledge, which can be expressed in words, numbers, scientific procedures and universal principles—seems to be as important as tacit knowledge—perceived as the unarticulated form of knowledge, based on human practices, as well as the ability to communicate and give personal representations of the world (Amin and Cohedent 2004). However, how are both explicit and tacit forms of knowledge be understood within the context of the agro-food economy and the knowledge economy? Is there a prioritisation between such forms of knowledge, or are they equally important for the production of new knowledge?

KNOWLEDGES AND KNOWLEDGE PRODUCTION IN THE KNOWLEDGE-BASED ECONOMY

As discussed, the term 'knowledge economy' could be seen as a new 'case study' of capitalism that is based on the appropriation of non-commodified material and immaterial processes and their transformation into objects for capital accumulation. Building upon a specific master economic narrative—descriptive of a situation but also prescriptive of the kinds of futures to be attained—as well as a strategic promissory vision for future development of economies and societies, it becomes a self-fulfilling prophecy which succeeds in shaping current socio-economic realities in its own image (Merton 1948; Sunder Rajan 2006; Ponte 2009; Jasanoff et al. 2008). In this way, it constitutes an *economic imaginary* (Jessop 2008) which also goes beyond the remits of an 'imaginary', as it succeeds in becoming real, constituting an economic narrative that can lead to the consolidation of an accumulation regime and the articulation of a specific regulatory mode that is capable of organising and structuring economies and societies in ways that can go along the articulation of a grand narrative and strategic vision while marginalising other, possibly better, directions of development.

However, within the KBE, what is the role of knowledge in meeting the economic goal of post-Fordist accumulation strategies? As also stated in the previous chapter, knowledge needed to be translated in a narrowly defined technocratic way, which would prioritise the urgency of high-tech innovation. In turn, 'innovation' was understood as 'the transformation of knowledge into novel wealth-creating technologies, products and services' (Asheim and Coenen 2006, p. 149). Within the OECD and EU context, and under the pressure of competition against rising economies in an increasingly globalised economic arena, innovation was perceived as prerequisite for future economic development. As Birch identified (2007), in the KBE narrative, innovation has most of the times been perceived in terms of revolutions in technology, such as the 'biotechnology revolution', which could in turn lead to dramatic changes in economies through the empowerment of existing technological regimes and their promissory visions for future economic growth. However, to what extent is this true? How is innovation taking place within the knowledge economy? Does innovation always relate to or stem from technological developments? Does 'innovation' in the 'knowledge economy' lie purely in explicit forms of knowledge?

In its essence, the spark of innovation requires the interplay between different types of knowledge. Smith (2000) underlined the way that modern innovation does not rest so much on discovery, but on learning, which does not necessarily imply the discovery of new technical or scientific principles but can also be equally based on the recombination or adaptation of existing forms of knowledge. In this context, innovation primarily constitutes a learning process, based on a complex web of relations, networks and knowledges (e.g. Birch 2007; Amin and Cohedent 2004; Nonaka 1994; Blackler 1995). As also acknowledged by OECD (1996), it is the result of numerous interactions within a community of actors and institutions—varying from firms and laboratories to academic institutions and consumers. Such understandings are important to realise that innovation is not just associated with detached codified knowledge, but is the outcome of a continuous dialogue between tacit and explicit forms of knowledge (Amin and Cohedent 2004; Birch 2007). In doing so, they are also important for going beyond a narrow 'knowledge-as-information' approach (Arrow 1962; Machlup 1980).

Nonaka and Takeuchi (1995), Blackler (1995) and Lam (2000) have all provided typologies of the knowledge forms that are important for the creation of new knowledge within the context of the knowledge-based economy (KBE). For them, depending on the particular locations in which knowledge production takes place, four distinct knowledges—embrained, embodied, embedded and encoded (Collins 1993 in Blackler 1995)—co-exist and become important for the successful utilisation of existing knowledges as well as the creation of new knowledges:

(a) 'Embrained knowledge'—abstract knowledge based on the process of 'knowing that' (Ryles 1949) or 'knowing about' (James 1950). This is an individual and explicit form of knowledge, based on the individual's conceptual skills and cognitive 'higher level' abilities, as well as their rational understanding and knowledge of universal principles or laws (Lam 2000). It is formal, abstract and theoretical knowledge, learned through reading books and formal education (Durbin 2006). It is the 'know what' type of knowledge, which mainly consists of 'knowledge of the facts' based on 'information' (Lundvall and Johnson 1994; Morgan and Murdoch 2000).

(b) 'Embodied knowledge'. This is individual and tacit knowledge based on practical thinking and bodily experiences ('doing') rather than on abstract rules. It is context-specific, 'intimate knowledge', learned through experience and training based on apprenticeship relations and surfacing 'in light of a problem at hand' (Lam 2000). It is a process of knowing dependent on direct interactions between people and things. It resembles the 'know how' form (according to Lundvall and Johnson's (1994) typology in Morgan and Murdoch 2000), which, although initially used in a firm context, it also refers to the skills that can also flourish within a wider context of organisations.

(c) 'Encoded knowledge' is a decontextualised form of knowledge, usually conveyed by signs and symbols and stored in a set of written rules (Poster 1990; Blackler 1995; Lam 2000). It is a collective and standardised form of knowledge that aims to render tacit knowledge as explicitly as possible, through the use of language—for example, by codifying worker experiences and skills, personal observations and questions into objective scientific knowledge (Durbin 2006). It broadly corresponds with what Lundvall and Johnson call 'know why' knowledge (1994 in Morgan and Murdoch 2000), where knowledge is articulated in a body of objective principles and laws, mainly used by specialised organisations, including universities.

(d) 'Embedded knowledge' is a collective form of tacit knowledge, built into unwritten routines, habits and norms. It is usually produced within the context of social interactions within 'communities of practice' based on shared beliefs, cultural norms and understandings. It is characterised by socially interactive processes of knowing ('learning'), where knowledge remains dispersed and relation-specific (Brown and Duguid 1991; Lam 2000).[2] This form of knowledge relates to the 'know who' type of knowledge, and is increasingly dependent on the social skills of both individuals and organisations in order to gain access to others' 'know-how'.

This complex knowledge typology also provides clues to the more complex and diverse webs and processes of knowledge that are also important for the production of knowledge within the 'knowledge economy'. Despite the prevalence of codified knowledge within the dominant KBE narrative and vision, different knowledge types, both collective and individual, co-exist and interchange, also through a series of knowledge-conversion processes.

Nonaka (1994) has underlined the centrality of knowledge-conversion processes for the production of new knowledge within knowledge economies. He has provided a four-step model, which also highlights the centrality of both social spaces[3] and social interactions for the production of knowledge. First, at the stage of *socialisation*, knowledge is primarily tacit and can be acquired through a sharing of experiences between individuals. *Externalisation* is the second stage of knowledge conversion, where dialogue or collective reflection become important for the articulation of tacit knowledge in the form of explicit concepts, and its broader communication through the use of symbols. At the third stage of *combination*, knowledge becomes standardised and organised into a knowledge system. Finally, the familiarisation with the new knowledge through constant

---

[2] In this context, as Blacker claims, a re-stabilisation between the individual's skills and the interpersonal organisation skills, which finally leads to the creation of the 'organisational routine practices', is achieved in the absence of written rules (Blackler 1995; Lam 2000).

[3] Nonaka and Konno (1998) have underlined the importance of space ('ba' in the Japanese philosophical context) for the knowledge creation through different forms of engagement. According to their approach, depending on its different forms, 'ba' can constitute the spatial territory in which new knowledge could be produced through a variety of processes which could facilitate the dialectical relationship between tacit and explicit forms of knowledge.

learning-by-doing leads to the *internalisation* of the newly produced explicit knowledge and its transformation into embodied actions.

This framework also encourages to consider the more complex relations of interdependence and processes of interchange between explicit and tacit forms of knowledge, as well as understand the significance of social relations and interactions for the production of new knowledge and innovation. This is also acknowledged by OECD (1996), which, as back as 2000, grounded successful innovation in the emergence of new forms of 'collective social learning' (p. 25). In this case, both 'creative activity', and interaction are considered important for the production of knowledge, which is thus configured as part of such embodied social practices (Amin and Cohedent 2004). They are the product of working together in 'communities of practice' (Knorr Cetina 1981) which come to produce new knowledge from their routine activities and make learning an unintended consequence of information sharing, collaboration and narration.[4,5] Following Callon (1999) and Latour (1986), innovation constitutes the product of connections between previously unconnected and heterogeneous ingredients, and an outcome of learning-through-doing and social engagement acquired within routines, conversations, meetings, scripts, memories and stories. But also, following Amin and Cohedent (2004), it can also be located in different 'spaces of knowing', where socio-spatial ties, relational or social proximity within a proximate or distant territory also become important for the production of new knowledge.

All these observations encourage us to re-think the knowledge social dynamics within the KBE and their centrality in the transformation of the KBE into a master economic narrative, which succeeds in constituting part of current socio-economic realities as well as shaping realities in its own image. Knowledge production appears to be a knowing and learning process that crucially depends on the co-existence of the different knowledge types, diverse social relations and knowledge conversion processes. As outlined, explicit and tacit forms of knowledge seem to complement each

---

[4] Orr (Amin and Cohedent 2004) has provided an example with regard to knowledge production in Xerox's highly skilled technical workers. As he found out, breakfast and lunch have been significant moments for workers to come together and learn through discussion of common problems, narration of experiences, joint problem solving and reflexivity through interaction.

[5] The communities of practice can be opposed to Lave and Wenger's conceptualisation of epistemic communities (Amin and Cohedent 2004) characterised by their heterogeneity, procedural organisation and focus on explicit knowledge.

other in the utilisation and production of new knowledge and innovation. 'Knowledge spaces' are configured as 'social spaces' that not only facilitate the articulation and communication of knowledge, but also they contribute to the production of new knowledge through communication. In this context, knowledge becomes the outcome of complex knowledge-conversion processes based on a combination of experiential understandings and tacit knowledges. Such processes challenge traditional understandings of knowledge production, the dominant role of 'epistemic communities' and explicit forms of knowledge within them, reinforcing the significance of a wider spectrum of communities, actors and knowledges for the production of innovation.

Such observations encourage us to not only reconsider scientific knowledge as a form of knowledge, but also re-think the role of scientific knowledge for the production of new knowledge and innovation within the KBE. They encourage us to not only re-think the role of tacit forms of knowledge for the production of innovation, but also consider the more complex web of knowledges for the production of innovation within the KBE. By doing so, they also encourage us to challenge dominant techno-scientific understandings of innovation, as well as approach innovation as a knowing processes based on the interchange of diverse knowledge patterns and interactions. However, such observations also provide the grounds for exploring the processes of production of knowledge and innovation within an agro-food economy, and specifically explore: What is the role of explicit and tacit forms of knowledge within the agro-food economy? What is perceived as innovation within the agro-food sector, and how is it conceived as taking place? Could, in this sense, the agro-food economy constitute a knowledge economy? In which ways?

KNOWLEDGE AND KNOWLEDGE PRODUCTION IN AGRICULTURAL SYSTEMS

Without doubt, agriculture constitutes one of the most important economic and employment sectors worldwide (Grigg 1982; Newby 1987; Newby et al. 1978). However, as Lenin stressed in his *Theory of the Agrarian question*,[6] the natural and socio-spatial specificities involved in

---

[6] 'Agriculture possesses certain peculiar features which cannot possibly be removed (if we leave aside the extremely remote and extremely problematical possibility of producing albumen and foods by artificial processes)' (Lenin 1938, p. 85).

the historical processes of agriculture and its connectedness to the 'laws of nature' have provided obstacles to its seamless incorporation into capitalist modes of production (Fine and Leopold 1994). Even within the more recent 'knowledge turn' of capitalist economies, agriculture is often assumed to maintain some traditional characteristics that separate them from a 'knowledge economy' (see Morgan and Murdoch 2000). However, inarguably, knowledge is an important factor in the historic and economic development of agriculture. So, what can this tell us about re-thinking the agro-food economy as a knowledge economy?

Knowledge in Pre-industrial Agriculture: The Dominance of Tacit Knowledge

Looking at the history of agriculture, knowledge—both in its tacit and explicit forms—had a central role to play, as an obstacle but also as an enabler of capitalist relations of production (Tribe 1981). Most agro-food studies underline the significance of tacit forms of knowledge in traditional agricultural processes. It was the 'art de la localité' (Mendras 1970), a specific knowledge system based on craft tasks and a constant engagement with land and the seed. Knowledge production within traditional agro-food systems was frequently interwoven with a dynamic labour process involving the 'savoir-faire paysan' (Lacroix 1981, p. 95 in van der Ploeg 1993), a labour process developed under constant interaction between people and nature. Tacit knowledge was a 'from practice to practice' knowledge (Bourdieu 1990), trespassing the stage of 'discourse' and 'theory'. It was a knowledge system based on the personal knowledge of the field, the genetic stock and the folk taxonomy that the 'knowledgeable actor', that is the 'agent' of the integrated manual and mental work, had at her disposal (Mendras 1970). However, it was also a knowledge system that was not unrelated to science (van der Ploeg 1993); it was just 'organised in a way that markedly differ[ed] from scientific discourse' (ibid., p. 210).

'Local', 'traditional', 'indigenous or ethnic', 'lay', 'non-expert' or 'non-professional knowledge' are some of the terms attributed to non-explicit forms of knowledge relating to agriculture (Tovey 2008). Following Latour's understanding of scientific knowledge as 'immutable mobile', tacit knowledge is the 'mutable immobile'[7] (Kloppenburg 1991; Tovey 2008) socio-spatially bounded knowledge, a personal and context-dependent knowledge based on personal interactions and shared experiences. Thus, agricultural practices were based on types of 'situated' or 'embodied' knowledge, which were based on the knowledge workers' personal engagement with land and nature; they constituted the 'knowledge practices', embodied and materialised not only by humans but also by hybrid collectives of both humans and non-humans (Ingold 2000; Haraway 1991). In this context, knowledge was always objective. It was embedded within the continuum of social, cultural and practical life and a direct connection with nature (Van der Ploeg 1993).

Knowledge in Industrial Agriculture: The Emergence of Science

With the advent of industrialised agriculture, tacit knowledge came to constitute a barrier to the incorporation of agriculture into the capitalist mode of production. In this case, scientific knowledge was key mechanism towards the 'rationalisation of farming processes' through a gradual detachment of knowledge from certain social and natural environments (Tribe 1978). New mechanical tools with regard to ploughing and tillage came to signal a new phase of human/nature divide, in which humans were conceived as the 'masters' and exploiters of nature. Traditional horse-powered ploughing was gradually replaced by heavy plough, led by new steam-powered machines (e.g. engines and tractors). Knowledge about specific chemical substances (e.g. phosphorus, potassium, nitrogen) was also key for the introduction of the new stage of a 'chemical', also widely known as 'green', revolution in agriculture, accompanied by the discovery of synthetic fertilisers, antibiotics and vaccines, synthetic pesticides. Darwin's theory of evolution, the pure-line theory of Johannson, and the re-discovery of Mendel's law of genetic inheritance were pivotal for the rise of plant breeding that would favour specific traits and create new breeds with improved productivity and survival rates. They also constituted harbingers for a greater genetic transformation in the mid twentieth century, based on Watson and Crick's earlier discovery on the DNA structure (Pardey and Beintema 2001). As Kloppenburg describes (1988), seed, which used to constitute a biological barrier to the industrialisation

---

[7] For Tovey, it is the sort of knowledge 'that remains particularistic and oriented to understand in detail a particular place and context' (2008, p. 194).

of agriculture, could now work as a locus of industrial appropriation. Thus, with the new biochemical and molecular genetics of the 1950s and the 1960s, new technologies were placed at the heart of the new agricultural development model based on agricultural science (Ward 1994 in Morgan and Murdoch 2000). Technological innovation became a key factor for an increase in output and productivity through the manipulation of the seed, the alteration of traditional agricultural practices and the displacement of the farmers' tacit agricultural knowledge.

In this way, a complex re-organisation of the agro-food system, as well as the farming routines, knowledge and practices was inevitable (van der Ploeg 1993). The agro-food economy expanded to include a wider spectrum of 'knowledgeable actors'. Primary agriculture became integrated into a broader 'agri-industrial complex', where large industrial units could operate both upstream and downstream of the farm (Goodman and Redclift 1991). Major innovations in production, processing and retailing came into play, transforming agriculture and food into a knowledge-based economic sector expanding beyond farm-based knowledge practices (Morgan and Murdoch 2000). Traditional farming practices, techniques of plant breeding, crop rotations and ploughing shifted from a craft to a truly scientific basis, attached to emerging innovations. In this way, as van der Ploeg argues (1993), 'agriculture was disconnected from those structuring elements that initially introduced specificity to it' (p. 21). Science appeared to contribute to a radical re-making of nature (Goodman and Redclift 1991). Thus farmers' tacit knowledges, traditionally attuned to the rhythms of nature and their personal relationship with local ecosystems, were gradually replaced by standardised knowledges, practices and procedures, frequently shaped around their newly established relations with and dependencies on large powerful agrifood institutions and stakeholders (Morgan and Murdoch 2000).[8]

This new agrifood model not only had an impact on farmers' traditional knowledges and farming practices. As already discussed, agriculture was increasingly entering into a broader 'agri-industrial complex' (Goodman and Redclift 1991) which could move beyond purely on-farm practices.

---

[8] Goodman et al. (1987) identify two principles in which the industrialisation of agriculture has taken place. First, appropriationism, in which integral elements to the agricultural production process are extracted and transformed into industrial activities; and second, substitutionism, in which agricultural products are reduced first to an industrial input and increasingly replaced by manufactured non-agricultural components.

Farmers were increasingly socio-economically dependent and bounded to new industry stakeholders related to agrifood processing and retailing. Newby et al. (1978) provided an interesting typology of farmers, which highlights the gradual incorporation of new technical knowledges and skills within farming communities, sometimes also resulting in a greater diversification within farming communities and a transformation of part of these communities into financial managers and entrepreneurs. In their survey—focusing on mid-1970s East Anglia—Newby et al. (1978) identify four different types of farmers—namely the family farmers, the gentleman farmers, the active managerial farmers and the agri-businessmen—illustrating such differences, and the gradual bureaucratisation of farming through the incorporation of new market interests and skills. However this also varied depending on the motivations and values of the different farming communities. For example, family farmers maintained a personal, long-lasting connection to land, shaping largely their everyday knowledge and farming practices. This significantly differentiated them from gentlemen farmers, who, despite their diachronic links with their land-holdings, they usually chose to distance themselves from everyday agricultural practices. Also, although for both farmer categories, 'profitability' (ibid., p. 182) was a key motivation, for the 'active managerial' and 'agri-businessmen' farmers, profit-maximisation was pivotal in shaping the increasing diversity of the off-farm, entrepreneurial knowledges and skills acquired by these categories of farmers. As Davis and Hinshaw (1957) put it, farmers were increasingly appearing in a business suit. In many cases, agri-businessmen were mainly 'business entrepreneurs who happened to be in the business of growing crops and rearing animals' (Newby et al. 1978, p. 181), thus signalling a new phase of farming where administrative, financial and accounting skills, had started being increasingly important for what was considered as the future of the agriculture and food.

This, of course, also reveals that agriculture has traditionally been a knowledge-based economic practice, where different knowledge forms, both tacit and explicit, seemed to compete or inhabit separate agro-food spaces, which would focus or prioritise certain types of agricultural practices. However, it seems that, gradually, tacit knowledges started losing their significance, resulting to a level of standardisation of farmers' local knowledge practices. However, based on our understanding of 'knowledge-based economy' as a complex interchange between both tacit and explicit forms of knowledge, could we really speak about a clear separation between tacit and explicit forms of knowledge in agriculture? And what would this mean in terms of our understanding of an agro-food economy as a knowledge economy?

Knowledge and Late Industrial Agriculture: Towards a Co-existence of Tacit and Explicit Knowledge

The 1970s and 1980s signalled the rise of a new 'food regime' incorporating a broader range of practices, knowledges and skills attached to both globalised industrial agriculture as well as alternative agro-ecological practices. In this context, the authority of science for the production of agrifood innovation was questioned, whereas farmers' knowledge regained its significance in the production of future agrifood innovation (Stuiver et al. 2004). However, as Stuiver et al. (2004) underlined, a recombination between explicit and tacit knowledges is also inevitable: science always exists but also gets meshed with farmers' knowledge of the farm, the animals, the soil. Thus, scientific knowledge, as the form of 'universal knowledge', gets localised to the farmer's specific setting' and personal experiences of the natural phenomena. But also, farmers come to develop their own language and classification regimes describing diverse categories of plants, land, soil and natural resources.

The rising number of food scares and crises raised levels of public mistrust towards science-based agricultural systems. Science's authoritative claim to truth has been challenged and the notion of 'expertise' started to encompass a wider number of non-accredited (Tovey 2008) or uncertified experts. In this context, farmers' knowledge got re-valorised, whereas new hybrid agricultural knowledge systems emerged, based on a complex combination of different forms of tacit knowledge, such as traditional, lay, non-expert or new 'local expert'. Following Funtowicz and Ravetz's understanding of a 'post-normal science' (1994a), we could claim that the hegemony of the traditionally perceived as 'normal' (in the Kuhnian sense) science[9]—over all other ways of knowing has been replaced by a more complex knowledge system based on a diversity of experts (see also Funtowicz and Ravetz 1994a,

---

[9] 'Normal science' here refers to knowledge based on reductionist mathematical explanations and observations made by a detached observer reproducing a fact/value distinction.

1994b). In this context, traditional knowledge[10] is configured as a 'local expert knowledge' system, which, compared to scientific knowledge, is neither standardised nor formal. It is the technical knowledge acquired by 'local experts'—farmers and producers—through experiential understandings with certain agro-ecological contexts (Collins and Evans, 2002). It constitutes a product of 'experience-based experts' or 'uncertified experts', equipped with a variety of tacit knowledges and skills that are important not only in the formulation of explicit forms of knowledge but also in the production of new knowledge.

In this framework, agricultural knowledges are 'situated knowledges' (Haraway 1991; Ingold 2000). They become important in enacting a science-based knowledge system that goes beyond the wandering eye of masculine scientific objectivity, in order to become more responsive to challenges, knowledges and values of diverse cultures and geographies of food. Following Antweiler (1998), we could claim that, in the context of knowledge practices of late industrial agriculture, local knowledge is not only referring to a specific location, but, as Antweiler (1998) identifies, it is culturally and ecologically situated. It is the type of knowledge which should not be contrasted with science (Sillitoe 2006), but builds on a continuum between formal science and everyday rationality (Antweiler 1998). It expands the limits of technological or environmental knowledge in order to incorporate knowledge of the social environments, based on complex social interactions and exchanges of knowledge (Antweiler 1998). Agricultural knowledge is therefore a hybrid type of knowledge, where traditional and modern (or post-modern), local and global are becoming intrinsically mingled together, created by complicated life-worlds. As Nygren identifies (1999), it is the form of knowledge where the 'boundaries between people's science and scientists' science' (p. 282) become impossible to draw. It is in this context that scientific knowledge is no longer configured as a system of abstract statements or universal truths. It is deeply heterogeneous, based on a diversity of components—people, skills, local knowledge and equipment—coming together for the enactment of a new scientific knowledge that can transcend its geographical and temporal boundaries of production (Turnbull and Verran 1995), drawing

---

[10] Following Nygren (1999), traditional knowledge is seen as a set of common practices collectively held and transferred from a generation to generation, based on their cultural values, their particular belief and knowledge systems.

on a variety of partial truths and disconnected components coming from an equally diverse set of times, places and circumstances.

The Case of Organic Agriculture and Beyond

Organic agriculture is an interesting example of such mingling of explicit and tacit forms of knowledges. Introducing a radical discontinuity from the productivist agro-industrial paradigm, organic agriculture aims to build and re-evaluate a more diverse spectrum of knowledges (Morgan and Murdoch 2000). Also described as 'radical innovation (Dosi et al. 1993), it has been seen as the main alternative to the conventional agrofood knowledge system, requiring innovators to forget much of their previously acquired knowledge (Johnson 1992). It underlines the significance of farmers tacit knowledges as well as relations of peer-to-peer knowledge sharing. It has also been described as the agricultural model for the resurrection of local, context-dependent knowledges, but also a complex agroecological knowledge system based on complex interactions between 'local experts' as well as 'external' science-based experts, coming together in creative ways in order to develop an innovative 'learning system' where a combination of farmers' local or indigenous knowledge and science-based knowledge is possible (Bager and Proost 1997).

A combination of tacit and explicit forms of knowledge is characteristic of other examples of alternative agro-food networks. AAFNs have been claimed to re-appropriate a diverse set of knowledges, varying from artisanal production techniques to marketing methods and social interaction modes between producers and consumers, In this context, leadership and professional management skills, entrepreneurial skills, strategic coaching and consumer learning have all been identified as important knowledges and skills to be acquired (Knickel et al. 2008; Torjusen et al. 2008). In this study, Fonte and Grando (2006, p. 44) attempted to map the different knowledges that come together within local and organic agro-food systems, thus encouraging us to consider and develop a more inclusive definition of expertise, based on such complex combinations between old and new, traditional and technocratic, tacit and codified forms of knowledge. As they describe, organic farming is configured as the knowledge system, which not only brings local and expert organic knowledge practices together. It is also a knowledge system which moves beyond farm-based knowledge practices, to incorporate managerial and technical knowledges and skills based on the involvement of a broader spectrum of knowledgeable actors, including certification bodies and organic institutions.

Such complex knowledge combinations can also be pivotal in understanding debates over the conventionalisation and bifurcation of organic agriculture (Goodman and Goodman 2007). In response to that, new agricultural knowledge spaces, also described as 'post-organic' or 'localist', have also emerged (Moore 2004). Such movements also move beyond the technical/practical knowledges involved in agro-food production to also consider knowledge production processes related to distribution and other relations of exchange. Aiming to challenge relations of socio-spatial separation and 'time-space distanciation' (Allen et al. 2003, p. 73; Buttel 2005), they underline the significance of 'know who', the communicative and social skills in configuring the agro-food spaces as knowledge spaces that are based on a more direct connection and trust-based relation between producers and consumers. They thus become important in not only realising the more complex forms of knowledges, both tacit and explicit, that exist within the alternative agro-food sector, not only realising the more diverse spectrum of practices, both on-farm and off-farm, that such knowledges are for, but also the broader spectrum of knowledges—for example both farm-based and communicative—that constitute important elements of it. It underlines the significance of both knowledge and mobilisation of knowledge (Tovey 2008), but it also helps us understand the innovative character of such practices, and realise the blurred boundaries between 'local innovation' and 'scientific innovation'.

Intermediary Summary: Re-thinking Knowledge Production in Agriculture

The above analysis allows us to appreciate the centrality of knowledge in configuring economies around agriculture and food, especially by moving beyond binary distinctions between expert and lay, scientists' and farmers' knowledges. Thus, although, historically, especially over the course of industrialisation of agriculture, a prioritisation of scientific knowledge appeared to resonate with the prevailing calculative logic of emerging agro-industrial economies, a co-existence of tacit and explicit forms of knowledge has also been inevitable. Despite the dominating role of science, the increasing scienticisation of nature and the subsequent divides between nature and society, new hybrid systems of knowledge emerge, based on complex interchanges between scientific and locally specific ways of knowing, developed in response and with respect to global and local environmental challenges (Jasanoff and Martello 2004, p. 335). In this context, 'co-producing' knowledge between a greater diversity of 'experts' becomes important in understanding agricultural societies as 'learning societies' (Buckmeier and Tovey 2008), or else new 'spaces of knowing' based on a constant exchange of knowledges as well as perceptions around natural environments. This also pivots a re-conceptualisation of the notion of 'expertise', which needs to expand to not only include scientific and bureaucratic-managerial experts but also experts trained in specific traditional and artisan modes and particular localities. Just like in Epstein's (1995) study of the central role of AIDS patients and their knowledge in shaping medical research on AIDS, in agriculture, such understandings are important in challenging the idea of 'experts' and 'expertise' in agriculture and food, and appreciating the role of a more diverse spectrum of stakeholders as well as lay publics, their knowledges and viewpoints in informing as well as shaping knowledge production processes, as well as influencing and challenging what constitutes science, what it is for or it should do.

Alternative agro-food systems provide an interesting knowledge space that signals and necessitates a shift towards a more dialectic relationship between different types of knowledge as well as a greater mobilisation of knowledge across space and time (Tovey 2008). In this context, the boundaries between different knowledge patterns and systems are vague and indefinable, as knowledge is reinvigorated as an entity that is both local and global (Turnbull and Verran 1995). A co-existence and combination of diverse knowledge forms (scientific and local, science expert and local expert, explicit and tacit, technical and social) become significant in the configuration of a new 'co-produced' knowledge model, in which, as Polanyi identified, all knowledge can be understood either as tacit or as rooted in tacit knowledges (1967). Science appears to no longer hold the sole authoritative claim to truth that it was once credited with, and what counts as 'expertise' is increasingly contextualised and located at the situation of its construction and application (Haraway 1991; Tovey 2008). New knowledge models and skills, as well as the diverse values embedded in them, appear significant in the re-making of agro-food societies as knowledge societies in ways that would also break the divide between powerful and powerless (Turnbull and Verran 1995).

However, what are the similarities and differences between the KBE and contemporary agro-food economies? Could the latter constitute a knowledge-based economy and on what grounds? And, what would this imply for our investigation of the AAFNs as a KBBE?

The Agro-food Economy as a Knowledge Economy?

In this historical account of agriculture and food, knowledge seems to have played a key role in the configuration of the agro-food sector as an economic sector. As described above, explicit and tacit forms of knowledge seemed to co-exist and interchange, not only in the context of increasing agricultural productivity, but also for the production of new knowledge in the context of increasing productivity around agriculture and food. Thus, despite their complex materially entangled nature, agricultural practices seem to also be configured as knowledge practices, increasingly embedded in dematerialised commodities and processes of labour that are also characteristic of those economies conceived as 'knowledge economies'.

For example, in many ways, farmers can be understood as 'knowledge workers', constituting an immaterial form of labour: they use and appropriate various forms of knowledges and skills, which become integral in their complex socio-material entanglements with land and nature. As Hardt and Negri also describe (2004, p. 10), farmers have always used the knowledge, intelligence and innovation typical of immaterial labour. A permanent interaction between mental and manual labour is also central to the accomplishment of their practices. Their work on and with the land has always been the product of a wider mental process of interpretation and evaluation of the process of production (van der Ploeg 1993). Thus, on the one hand, farmers' labouring processes appear to not only be the outcome of specific technical knowledge, but also of the broader understanding of and engagement with the complex social and material environments around them. Conceived as 'knowledge workers', farmers seem to use and apply their existing knowledges and skills, as well as extract and acquire new knowledge from those broader socio-material environments in which they become embedded. As Hardt and Negri describe (Hardt and Negri 2004), although very physical, traditional agricultural work can be considered as science: for example, in the sense that every agriculturalist is also expected to perform the role of a chemist—for example, in the case of matching soil types and transforming fruit and milk into wine and cheese—or the role a genetic biologist—for example, selecting the best seeds to improve plant varieties—or the role of a meteorologist—for example, watching the natural climatic phenomena.

However, a dematerialisation of labour processes is also evident within the more recent agro-industrial and techno-scientific developments in agriculture and food. The cultivation of new industrial or cash-crops, the use of chemical pesticides and machinery provide clues to the changing forms of both material and immaterial, manual and mental, forms of labour that came with such developments. Farmers' knowledge practices came to adapt to the changing farming socio-material environments, the new processes of standardisation and technological development that have been brought with it. In the course of time, farmers also had to embody a greater diversity of expertise related to the finance and marketing of their business. However, there has also been an increasing number of new off-farm knowledge workers that are also pivotal in understanding processes of dematerialisation of labour in agricultural practices. The engagement of plant and life scientists, economists, computer scientists in agricultural practices indicates not only the greater diversity of knowledges and expertise agriculture and food had been increasingly dependent on, but also the changing nature of those, originally conceived as traditional, farmers' knowledges and practices.

In this context, the agro-food economy appeared to also re-iterate the profit-oriented rationale of the dominant KBE narratives. In many ways, knowledge is perceived key at generating profit not only through enhancing agricultural productivity, but also through the production and commodification of this new knowledge that is promised to enhance agricultural productivity and the profits deriving from such processes. In this context, despite a prevailing prioritisation of scientific knowledge, discourses and understandings of innovation as a process of co-existence and interchange between explicit and tacit forms of knowledge also becomes evident (Amin and Cohedent 2004; Birch 2007; Smith 2000; Lam 2000). Knowledge production, as an important component of the knowledge economy sector, does not reveal itself as a static process of scientific intervention for the production of new knowledge. Similarly to the knowledge economy narratives, knowledge production is conceived as a continuous knowing process, where the recombination and adaptation of existing forms of knowledge, the internalisation and externalisation of different knowledge forms (Nonaka 1994) become significant for fulfilling a strongly economising logic of productivity for the different, both on-farm and off-farm, 'communities of practice' and 'knowledge workers' (see Knorr Cetina 1981; Brown and Duguid 1991; Amin and Cohedent 2004).

Thus, in many ways, agro-food economies consist of various elements which could make us speak of them as knowledge-based economies. As already discussed, the agro-food practices could traditionally be considered as knowledge practices, combining both tacit and explicit forms of knowledge and a constant interchange between manual and mental, material and immaterial, labour emerging out of a continuous interaction and engagement with land and nature (van der Ploeg 1993). In particular, with the incorporation of new off-farm practices and, therefore, a greater diversity of actors[11] in the agro-food system,[12] a co-existence between different forms of knowledge becomes even more prevalent. The agro-food economic space is configured as a new 'knowing space', where communities of actors with different knowledges and skills contribute to the production of the new knowledge for agriculture and food. Knowledge appears to overcome a binary logic which would separate tacit and explicit forms of knowledge. However, following the mainstream KBE logic, a science-driven innovation logic prevails, which fails to meet processes of 'co-production' between social and epistemic orders (Jasanoff et al. 2008). Thus, despite the increasing important role of tacit knowledge in informing and shaping processes of scientific innovation, a subordinate role of tacit to explicit forms of knowledge—usually perceived in the form of techno-scientific knowledge—prevails. With the contemporary agro-food economy, different communities of knowledge workers come to combine their knowledges and skills, therefore re-configuring the agro-food 'knowledge space' as a 'social space' of knowing. However, in many ways, like with most cases within the mainstream KBE sectors, such agrifood spaces appear to still be in need of a more equitable open dialogue and democratic processes of collaboration between different communities of experts, including those previously understood as non-experts.

Profitability is another important parameter in understanding the agro-food economy as a knowledge economy. Following the economising logic of the mainstream knowledge economy narrative, the production of knowledge within the emerging agro-food economies should not be

---

[11] For example, scientists with advanced knowledge of agro-ecology, biology and economics, as well as farmers with lay knowledge about land, natural processes and the soil.

[12] For example, as discussed above, agro-food practices related to new agro-food production methods, as well as retailing and marketing.

detached from ideas of productivity and profit. In the case of the agro-food sector, such ideas of productivity can vary. On the one hand, knowledge production within the agro-food sector is associated with an increase of agricultural productivity. In this particular context, agricultural productivity can refer to both human and non-human labour productivity—for example, the role of machinery in overcoming limitations of human agricultural labour, but also the role of new technologies—for example, genetic engineering—in enhancing nature's productivity. However, in the agro-food economic sector, profitability moves beyond the socio-material boundaries of on-farm practices and actors. Within the context of the agro-food knowledge economy, capital accumulation appears to also not only through control over the means of production, but also through commodification of knowledge about nature, as well as commodification of nature itself (see Brennan 2000; Thacker 2005). In this way, within the context of an agro-food knowledge economy, profit does not necessarily come from an increased labour productivity related to on-farm practices, but also from processes of commodification of knowledge, which are promised to contribute to an increased productivity of on-farm labour processes.

The above analysis provides clues to the ways an agro-food economy can be understood as a knowledge economy—for example, through the different, both explicit and tacit, forms of knowledges, the different, both material and immaterial, labour processes involved, the different communities of practice and certified and uncertified experts, as well as the way they interact and come together for the production of agrifood knowledge and innovation. Of course, as part of this, it is also important to understand the reproduction of injustices that are also emblematic of mainstream knowledge economies: the prioritisation of scientific vs. other forms of knowledge in the production of agrifood knowledge, the uneven processes of participation of experts for the production of agrifood innovation, and of course, its narrow profit-based promissory narrative based on commodification of both knowledge and life. However, what does this tell us about alternative agro-food networks and the knowledge economy? How would this help us think of an alternative agro-food economy as knowledge economy?

Grounds for the AAFNs as a KBBE: Some Preliminary Remarks

No doubt, in its essence, the alternative agro-food sector constitutes part of the agro-food economy. This is also evident in the divergent knowledges and knowledge production processes that also appear key in the configuration of an alternative agro-food system, such as the organic agricultural system discussed above.[13] However, as also discussed above, to a great extent, alternative agro-food networks (AAFNs) have also emerged as an alternative to the conventional productivist agro-food model. In this way, AAFNs have been considered to constitute an alternative to the dominant capitalist logic of contemporary agro-food economies—as also increasingly manifested in processes of commodification of knowledge and labour for agriculture and food. Thus, considering the centrality of knowledge as a resource, a commodity and a productive force, AAFNs could be understood as economic spaces that aim to challenge the knowledge-driven economising logic of conventional agro-food economies. And, considering the centrality of knowledge within them, they can also be understood as the knowing spaces, which, born out of such KBE-related agro-food developments, aim to also construct their own knowledge economic practices, which can be conceived as both part of and reaction to the mainstream KBE.

However, what does this mean, and how is it manifested in practice? What are the different knowledges and skills involved in AAFNs' agrifood knowledge production? How do tacit and explicit forms of knowledge come together in various social relations of production? Based on the knowledge production within the AAFN economy, what are the similarities and differences of this economy to the KB(B)E? Could their economy be considered as a knowledge economy? And, if so, what are the particular characteristics of their knowledge economy, and how does it relate to the mainstream KBBE? Are they just an alternative to the KBBE, or, could they also constitute an alternative KB(B)E?

Based on these questions, in the following chapters, I aim to shed light on the knowledge economic aspects that would help us conceptualise an AAFN economy as a knowledge economy. In particular, by looking at the knowledge and economic aspects of the AAFN economy, I further

---

[13] The latter has also been touched in the example of organic agriculture, which was largely posed as an alternative to the conventional, productivist agro-food model.

investigate their potential of AAFNs to constitute an alternative KB(B) E—possibly as a way of developing a different understanding of the latter. The first step in such an investigation focuses on the identification of the knowledge patterns and the knowledge production processes within the AAFN economy. However, before getting further with such an analysis, the following chapter also aims to offer an overview of the particular AAFNs as well as the research methods used for their investigation.

REFERENCES

Allen P., FitzSimmons M., Goodman M. and Warner K. (2003) Shifting plates in the agrifood landscape: the tectonics of alternative agrifood initiatives in California, *Journal of Rural Studies* 19(1): 61–75.

Amin, A. and Cohedent, P. (2004) *Architectures of Knowledge: Firms, Capabilities and Communities*. Oxford: Oxford University Press.

Antweiler, C. (1998) Local Knowledge and Local Knowing: An Anthropological analysis of contested "Cultural Products" in the Context of Development, *Anthropos* 93: 469–494.

Arrow, K. (1962) The Economic Implications of Learning by Doing. Review of Economic Studies, 29(3):155–173.

Asheim, BT and Coenen, L. (2006) Contextualising regional innovation systems in a globalising learning economy: on knowledge bases and institutional frameworks, *The Journal of Technology Transfer* 31:163–173.

Bager, T., Proost, J. (1997). Voluntary regulation and farmers' environmental behaviour in Denmark and The Netherlands, *Sociologia Ruralis* 37 (1): 79–98.

Birch, K. (2007) Knowledge, Place and Power: Conceptualising Value Creation in Knowledge-Based Commodity Chains. Working paper. Centre for Public Policy for Regions. University of Glasgow.

Blackler, F. (1995) Knowledge, Knowledge Work and Organizations: An Overview and Interpretation. *Organisation Studies*, 16(6): 1021–1046.

Bourdieu, P. (1990) *The Logic of Practice*. Cambridge UK: Polity Press: Oxford UK: B. Blackwell.

Brennan, T. (2000) *Exhausting Modernity: Grounds for a new economy*. London and New York: Routledge.

Brown, J.S. and Duguid, P. (1991) Organizational learning and communities of practice: towards a unified view of working, learning and innovation, *Organisation Science* 2(1):40–57.

Buttel, F. (2005) Ever Since High Tower: The Politics of Agricultural Research Activism in the Molecular Age, *Agriculture and Human Values* 22(3):275–283.

Callon, M. (1999) The Role of Lay People in the Production and Dissemination of Scientific Knowledge. *Science, Technology and Society* 4(1):81–94.

Collins, H. (1993) The structure of Knowledge, *Social Research*, 60:95–116.
Collins, H.M. and Evans, R. (2002) The Third Wave of Science Studies: Studies of Expertise and Experience, *Social Studies of Science* 32(2):235–296.
Davis, J. and Hinshaw, K. (1957) *Farmer in a Business Suit.* New York: Simon and Schuster, Inc.
Dosi, G., Faille, M. and Marengo, L. (1993) Organisational Capabilities, patterns of knowledge accumulation and governance structures in business firms: An Introduction, *Organisation Studies 29*: 1165–1185.
Dreyfus, H. L. and Rabinow, P. (1982) Michel Foucault: beyond Structuralism and Hermeneutics. University of Chicago.
Durbin, S. (2006) Who Gets to be a Knowledge Worker? The case of UK call centres. In Walby, S., Gottfried, H., Gottschall, K. & Osawa, M. (eds.) *Gendering the Knowledge Economy: Comparative Perspectives.* Palgrave Macmillan: pp. 228–249.
Epstein, St. (1995) The Construction of Lay Expertise: AIDS Activism and the Forging of Credibility in the Reform of Clinical Trials. Science, *Technology and Human Values* 20(4): 408–437.
Fine, B. and Leopold, I. (1994) *The World of Consumption.* London: Routledge.
Fonte, M. and Grando, S. (2006) A Local Habitation and a Name: Local Food and Knowledge Dynamics in Sustainable Rural Development, *CORASON project.* Available online at www.corason.hu.
Funtowicz, S.O. and Ravetz, J.R. (1994a) Uncertainty, Complexity and Post-Normal Science. *Environmental Toxicology and Chemistry* 13(12): 1981–1984.
Funtowicz, S.O. and Ravetz, J.R. (1994b) Emergent Complex Systems. *Futures* 26(6): 568–582.
Goodman, D. and Goodman, M. (2007) Localism, Livelihoods and the 'Post-Organic': Changing Perspectives on Alternative Food Networks in the United States. In Maye, D, Holloway, L. and Kneafsy, M. (eds.) *Alternative Food Geographies Representation and Practice.* Elsevier.
Goodman, M. and Redclift, M. (1991) *Refashioning Nature.* London: Routledge.
Goodman, D., Sorj, B. and Wilkinson, J. (1987) *From farming to biotechnology.* Oxford: Basil Blackwell.
Grigg, D. (1982) *The World Food Problem, 1950–80.* Oxford: Basil Blackwell.
Haraway, D. (1991) 'Situated knowledges: The science question in Feminism and the privilege of partial perspective', in Haraway, D. (ed.) *Simians, Cyborgs and Women: The Reinvention of Nature.* New York: Routledge.
Hardt, M. and Negri, A. (2004) *Multitude: War and Democracy at the Age of Empire.* New York: The Penguin Press.
Ingold, T. (2000) *The perception of the environment: essays on livelihood, dwelling and skill.* London: Routledge.
James, W. (1950) *The Principles of Psychology.* New York: Dover.
Jasanoff, S. and Martello, M.L. (2004) Conclusion: knowledge and governance. In S. Jasanoff and M.L. Martello (eds.), *Earthly politics: local and global in*

environmental governance Cambridge, MA: Massachusetts Institute of Technology. pp. 335–350.

Jasanoff, S., Kim, S-H. and Sperling, St. (2008) Socio-technical Imaginaries and Science and Technology Policy: A Cross-National Comparison. Project Summary.

Jessop, B. (2008) *State power: a strategic-relational approach.* Cambridge: Polity Press.

Johnson, B. (1992) Institutional learning. In Lundvall, B. (Ed.), *National Systems of Innovation.* London: Pinter.

Kloppenburg, J. (1988) *First The Seed: The Political Economy of Plant Biotechnology, 1492–2000.* Cambridge University Press.

Kloppenburg, J. Jr (1991) Social theory and the de/reconstruction of agricultural science: local knowledge for an alternative agriculture, *Rural Sociology* 56 (4): 519–548.

Knickel, K., Susanne von Münchhausen, Henk Renting and Sarah Peter (2008) Supporting collective action in alternative food networks: Findings from 18 in-depth case studies in ten European countries, Second International Working Conference for Social Scientists "Sustainable Consumption and Alternative Agri-Food Systems", 27–30 May 2008, Arlon. Available online at http://www.suscons.ulg.ac.be. Retrieved March 17, 2009.

Knorr Cetina, K. (1981) *The Manufacture of Knowledge.* Oxford: Pergamon Press.

Kuhn, T. (1962) *The Structure of Scientific Revolutions.* University of Chicago Press.

Lacroix, A. (1981) *Tranformations du process de travail agricole; indicidences de 'industrialisation sur les conditions de travail paysannes.* Grenoble: Institute National de la Recherche Agronomique.

Lam, A. (2000) Tacit Knowledge, Organisational learning and Societal Institutions: An Integrated Framework, *Organization Studies* 21(3):487–513.

Latour, B. (1986) Visualisation and Cognition: Thinking with Eyes and Hands, *Knowledge and Society* 6:1–40.

Lenin, V. I. (1938) *Theory of the Agrarian Question.* New York, NY: International Publishers, V. I. Lenin, Selected Works, Vol. XII.

Lundvall, B. and Johnson, B. (1994) The learning economy, *Journal of Industry Studies* 1(2): 23–42.

Machlup, F. (1980). *Knowledge and knowledge production. Knowledge: its creation, distribution, and economic significance. Vol. I.* Princeton, NJ: Princeton University Press.

Mendras, H. (1970) *The Vanishing Peasant: Innovation and Change in French Agriculture*, Cambridge: Cambridge University Press.

Merton, R.K. (1948) The Self-Fulfilling Prophecy, *The Antioch Review* 8(2): 193–210.

Moore, O. (2004) What Farmers' Markets Say about the Post-organic Movement in Ireland. In Holt, G.C. and Reed, M. (eds.) *Sociological perspectives of Organic Agriculture.* CAB International.

Morgan, K. and Murdoch, J. (2000) Organic versus Conventional Agriculture: Knowledge, Power and Innovation in the Food Chain, *Geoforum* 31:159–173.

Newby, H. (1987) *Country Life: A Social History of Rural England.* London: Weidenfeld and Nicolson.

Newby, H., Bell, C, Rose, D. and Saunders, P. (1978) *Property, Paternalism and Power: Class and Control in Rural England.* London, Hutchinson & Co.

Nonaka, I. (1994) A Dynamic Theory of Organisational Knowledge Creation, *Organisation Science* 5(1):14–37.

Nonaka, I. and Konno, N. (1998) The Concept of "Ba": Building a Foundation for Knowledge Creation, *California Management Review*, 40(3): 40–54.

Nonaka, I. and Takeuchi, H. (1995) *The Knowledge-Creating Company: How the Japanese Create the Dynamic Innovation.* New York: Oxford University Press.

Nowotny, H, Scott, P. and Gibbons, M. (2001) *Re-thinking science. Knowledge and the public in an age of uncertainty.* Cambridge: Polity Press.

Nygren, A. (1999) Local Knowledge in the Environment_Development Discourse: From dichotomies to situated knowledges, *Critique of Anthropology* 19:267–288.

OECD (1996) *The Knowledge-based Economy.* Paris: OECD.

Pardey, P.G. and Beintema, N.M. (2001) Slow Magic: Agricultural R&D A Century After Mendel. Agricultural Science and Technology Indicators Initiative. International Food Policy Research Institute. Washington, DC. October 2001. Available online at http://www.ifpri.org/sites/default/files/publications/fpr31.pdf.

Ploeg, J.D. van der (1993) Potatoes and Knowledge. In Hobart, M. (ed.) *An anthropological critique of development: The growth of Ignorance.* London and New York: Routledge: pp. 210–227.

Polanyi, M. (1967) *The Tacit Dimension.* New York: Anchor Books.

Ponte, S. (2009) From Fishery to Fork: Food Safety and Sustainability in the Knowledge-Based Bioeconomy, *Science as Culture* 18(4):483–495.

Poster, M. (1990) *The Mode of Information: Post-structuralism and Social Context.* Cambridge: Polity Press.

Ryles, G. (1949) *The Concept of Mind.* London: Hutchinson.

Sillitoe, P. (2006) Introduction: Indigenous Knowledge in Development. *Anthropology in Action* 13(3): 1–12.

Smith, K. (2000) What is the 'knowledge economy'? Knowledge-intensive industries and distributed knowledge bases. Paper presented to DRUID Summer Conference on The Learning Economy—Firms, Regions and Nation Specific Institutions. June 15–17, 2000.

Stuiver, M., Leeuwis, C. and van der Ploeg, J.D. (2004)The Power of Experience: Farmers' Knowledge and Sustainable Innovations in Agriculture. In J.S.C. Winskerke and J.D. van der Ploeg (eds.) *Seeds of Transition: Essays on novelty production, niches and regimes in agriculture*. The Netherlands: Koninkijke van Gorcum BV.

Sunder Rajan, K. (2006) *Biocapital: The Constitution of Postgenomic Life*. Durham and London: Duke University Press.

Thacker, E. (2005) *The Global Genome: Biotechnology, Politics and Culture*. Cambridge, MA: MIT Press.

Torjusen, H., Lieblein, G., Vittersø, G. (2008) Learning, communicating and eating in local food systems: the case of organic box schemes in Denmark and Norway, *Local Environment* 13(3): 219–234.

Tovey, H. (2008) Introduction: Rural Sustainable Development in a Knowledge Society Era, *Sociologia Ruralis* 48(3):185–199.

Tribe, K. (1978) *Land, Labour and Economic Discourse*. London: Routledge & Kegan Paul.

Tribe, K. (1981) *Genealogies of Capitalism*. London and Basingstoke, The Macmillan Press.

Turnbull, D. and Verran, S. (1995) Science and Other Indigenous Knowledges in Sheila Jasanoff et al. (eds.), *Handbook of Science and Technology Studies*, London/Thousand Oaks, Sage (revised edition 2002): pp. 115–139.

Ward, N. (1994) *Farming on the treadmill: agricultural change and pesticide pollution*. Unpublished PhD Thesis, University College London.

CHAPTER 4

Researching Alternative Agro-Food Networks in the Aftermath of the Knowledge Economy

This chapter is intended to provide an overview and justification of the case studies of AAFNs under investigation, that is, of the two AAFNs in two distinct counties in the Northwest of England: the rural county of Cumbria and the metropolitan region of Manchester. It starts by delving deeper into the explicit linkages of the British economy to the knowledge economy, but also underlining the centrality of the Northwest of England as a territory in which the agro-food economy is becoming key in the region's attempt to constitute a competitive knowledge economy, especially in the early 2000s. It then moves on to focusing on the networks under investigation, and provides details of the particular types, structural morphology and characteristics of the agro-food initiatives under investigation, also by situating them in the broader theoretical-methodological framework that has informed and shaped my research journey.

THE BRITISH AGRIFOOD LANDSCAPE IN THE AFTERMATH OF THE KNOWLEDGE ECONOMY

Since the inception of the Knowledge-based Economy in the early 2000s, knowledge, in the form of the knowledge-based industries and occupations, has constituted the backbone of the UK economy. According to OECD/Eurostat (see Brinkley 2006), following the United States and Germany, the UK is ranked as the third biggest economy amongst the G7 economies, with, as of 2005, 48% of its total employment being related to the knowledge economy. The British knowledge economy appears to be based mainly on knowledge-based services constructed around financial and market, as well as business, high tech, education and cultural services (Brinkley 2006). This has also led to a significant concentration of the UK's knowledge economy in urban conglomerates of the British South, off-setting developments in the North.

The bio-economy constitutes part of this service-based knowledge economy, and the UK considered as a leading country in the key areas of research and innovation underpinning the bioeconomy—including agriculture and food. The bioeconomy also signalled a unique opportunity for the regions of the North to overcome the emerging North/South knowledge economy divide. Thus, despite a prevailing concentration of the service-based economy in the South, the North of England has built a distinctive set of bioeconomy assets, acclaimed to be mainly positioned in the world-class science, applied research excellence and translational expertise, as well as an industrial capacity reaching the number of 16,000 companies in the region (Bauen et al. 2016). Agriculture and food have constituted fertile spaces for advancing the North's knowledge bioeconomy, and creating a competitive advantage for the region's knowledge economy. In this vision, agri-tech and industrial biotechnology—including advanced biofuels and bio-based jet fuels, novel foods, high-value chemicals from novel crops, functional foods and nutraceuticals, novel crop varieties that are resistant to pests, diseases and climate-related stresses—are envisioned key in 'achiev[ing] full exploitation of the bio-economy to deliver jobs and economic growth' for the North (Bauen et al. 2016). All these developments manifest the significance of the North as a region for the country's knowledge bioeconomy, and therefore, highlighting the significance of a further investigation of the possibilities for an alternative knowledge bioeconomy in the region.

With regard to agriculture, with the advent of the KBE, England appeared to be following its traditional, technologically driven productivist logic in its agricultural practices.[1] In the 2000s, conventional agriculture pertained as the dominant agrifood model in the UK, specifically accounting for the 99.7% of the farm land in England, with four multinational agro-chemical companies (ICI, Schering, Monsanto and Bayer),

---

[1] The thirty years from 1950 to 1980 constituted a 'chemical revolution' in British agriculture (p. 81), witnessing an eight-fold increase in pesticide sales between 1948 and 1982, reaching the amount of £542 million per year amounting to 31,000 tons of pesticides applied to 3.8 million hectares of agricultural land (Ward 1994).

controlling over 60% of the agro-chemical land (Morgan and Murdoch 2000). The salmonella incidents of 1988, the BSE or 'mad cow disease' scandal of 1992, the foot and mouth epidemic of 2001 have also exacerbated pressure on farmers to invest in and adopt new, productive technologies in the promise of increased yields, output and productivity (Morgan and Murdoch 2000). Whereas the 2001 Curry report has been pivotal for undermining the role of British farming for Britain's national economy, describing existing agricultural practices as 'dysfunctional' and 'unsustainable' (see PCFF 2002), and thus prioritising the preservation of the English countryside and the local wildlife as key objectives of the national agricultural policies. This put additional pressure on farmers, as they would no longer be subsidised for producing food, but for becoming stewards of the British countryside and its wildlife.[2]

Responding to such complex challenges for the British agrifood economic sector, science and technology have been pivotal for transforming the British agrifood landscape into a more complex economic space, or what we also tend to call a complex agrifood supply chain, which would help maintain a competitive advantage in the country's extensively knowledge-driven economy. This, of course, entailed the incorporation of an increasing number of on-farm and off-farm services—such as the emerging agrifood science, as well as food and drink manufacturing, wholesaling and retailing, and non-residential catering—thus reinforcing a gradual integration of the agro-food economy into the service-based, knowledge-driven logic of the knowledge economy. It is remarkable to notice that, according to DEFRA, already in 2010, although the UK agro-food sector was contributing £84.6 billion or 7.1% of the national market sector[3] gross value added (GVA)[4] and employing 3.06M people, farming

---

[2] According to the DEFRA (2010), almost 5 million (one quarter) hectares of agricultural land in the UK are used for growing crops. Cereals such as wheat, barley and oilseeds make up almost 80% of the total production; other arable crops, such as proteins and sugar beet make up 13%; horticulture (fruit, vegetables and ornamentals) make up 4%; and potatoes use 3% of the land. The agro-industrial mode of production remains dominant, since only 700,000 hectares of land are estimated to utilise organic methods of production.

[3] According to DEFRA's Food Statistics Pocketbook (2010), the market sector covers private non-financial corporations, private financial corporations, household and public corporations. It excludes government and non-profit institutions serving households.

[4] According to *OECD Glossary of Statistical Terms*, gross value added is 'the value of output less the value of intermediate consumption; it is a measure of the contribution to Gross Domestic Product (GDP) made by an individual producer, industry or sector' (OECD 2012).

was only amounting for a 9.4% contribution of the total GVA, employing 0.45M of those 3.06M employees in the agrifood economy.

At the same time, the bioeconomy appeared to constitute a significantly larger space of the agrifood economy. Agriculture and food constitute part of the UK's transformational bioeconomy,[5] which accounts for 3.5% of gross value added in the UK in 2014 (£56.0 billion). According to a BBSRC report, almost half of the employment in the bioeconomy is in agriculture and fisheries. Food and drink manufacturing account for almost 40% of the UK bioeconomy, with industrial biotechnology and bioenergy contributing to 13% (Bauen et al. 2016). Agri-tech businesses also have a significant presence in the UK, with the government having invested £68 M in three new Centres for Agricultural Innovation—covering livestock, crop protection, and engineering—to help translate agricultural innovation into commercial opportunities: Agrimetrics, a big data centre of excellence for the whole food system, and the Agri-Tech Catalyst, which was set up by the Department for Business, Innovation and Skills, Innovate UK, the Department for International Development and the BBSRC, to help businesses and researchers commercialise their research and develop innovative solutions to global challenges in the agriculture sector (Bauen et al. 2016).

In the North of England, food and drink represents around one-third of the regional bioeconomy. Organisations in the region are well represented in competitions such as the Industrial Biotechnology and Agri-Tech Catalysts, and in other Innovate UK funding competitions relevant to the agrifood bioeconomy. Much of the region's engagement in the agrifood bioeconomy derives from university-industry research collaborations on agriculture and food. Only between 2014 and 2017, projects with North of England partners won 35% of Agri-Tech and 29% of Industrial Biotechnology Catalyst grants. The N8 AgriFood, also known as the N8 Agri-Food Resilience Programme, is an example of such a research collaboration, which received £16 M for building a translational research platform promising to address food chain issues at local, national and international scales. The N8 AgriFood programme has an ambition to

---

[5] The whole bioeconomy, comprising transformative, upstream and downstream elements and induced effects, is a significant sector for the overall UK economy, generating approximately £220 billion in gross value added and supporting 5.2 million jobs in 2014. Agriculture and fishing, forestry and logging, water and remediation activities, food products and beverages constitute the UK's transformational bioeconomy, which amounts 3.5% of gross value added in the UK in 2014 (£56.0 billion).

ambition is to double the size of the transformative bioeconomy in the North of England in GVA terms from £12.5 billion now to £25 billion in 2030, not only by bringing Northern universities' to work closer together, but also by enabling industry to access as well as shape academic research in the areas of agriculture and food. As identified, 'Industrial Biotechnology for the Bioeconomy' constitutes a growing research community within the N8 universities, particularly specialising in the areas of agri-science and agri-tech.[6] Of course, such a research strategy is also based on a close collaboration with local and national agrifood industrial partners.

The North of England is also home to some of the largest bioeconomy innovation industrial centres. For example, Fera is a commercial business that sits between industry and government, supporting a wide range of agri-food innovations, including plant health and crop protection through food and feed safety and authenticity to novel biotechnologies. The National Agri-Food Innovation Campus (NAFIC) which hosts four centres, including the Centre for Innovation Excellence in Livestock and the Crop Health and Protection centre, and received £90 M government investment to improve technology uptake by UK companies in the agri-tech supply chain. The Food Innovation Network and the Stockbridge Technology Centre are also part of NAFIC, whereas BDC is another important translational organisation in the region, which as part of the BioPilots UK alliance of open-access biorefining centres, is aimed to develop ways to convert plants, microbes and biowastes into profitable biorenewable products.

However, as Lang et al. (2009) also identified, significant power shifts have also taken place within the food supply chain economy, suggesting also a move from a producer-driven to a consumer-driven economic model, and signalling the rise of food retailers as the dominant players in the agrifood supply chain. Alongside the rise of the service-based knowledge economy in the UK, four food retailers (Tesco, Sainsbury, Asda and Morrisons) had three quarters of all sales of food in the country, with one (Tesco) accounting for one-third of national sales. Such developments have also had significant implications not only in the nature and structure

---

[6] These include research on biotechnological processes for the conversion of varied wastes and cultivated feedstocks into higher-value ingredients for consumer products, the use of microbes and enzymes in fermentation and anaerobic digestion, catalytic and biocatalytic conversion of biomass, and microwave and hydrothermal treatment of lignocellulosic materials.

of the UK supply chain—with a decline in the number of small-scale local grocers and independent food shops (Thanassoulis 2009; DEFRA 2010; Blythman 2004)—but also, in the broader British food consumer—both eating and dining—culture. According to a survey (Mintel 2001 in Huxley 2003), by the year 2001, three quarters of British families were reported to have abandoned regular meals and one in five never sat together to eat. Convenience food, acquired in supermarkets or fast-food shops and restaurants, constituted a significant part of the British population's everyday dietary habits. Obesity and malnutrition emerged as the new food epidemics, while household food insecurity and food poverty emerged as public health issues, also related to an increasing number of food banks and food deserts.[7]

The 2000s also saw the emergence and rise of food banks across the country, rising from 41,000 in 2009 to 1.3 million packages of emergency food supplies in 2017–2018. Between 2007 and 2016, there has also been a 30% increase in malnutrition-related deaths in hospitals. As also explained in detail in the section below, in parallel to the rise of the service-driven knowledge economy, there is an exacerbation of food deserts and food poverty across the North of England, primarily affecting the increasing number of low-paid or unemployed part of the what is considered 'low-skilled' part of the population. An increasing number of children now living in conditions of household food insecurity—reaching the number of 132,000 children in the North East of England. Such phenomena come in direct contrast to the country's knowledge economy agenda for competitiveness, employment and growth. However, it is also representative of this new knowledge-driven economic order, highlighting the urgency for identifying the possibilities of an alternative agro-food knowledge economy, or even which would also help us re-think or even reclaim the latter.

Therefore, the North of England is an interesting, albeit contradictory, agrifood landscape, which further underlines the urgency for reclaiming the dominant agrifood knowledge economy. On the one hand, as we have seen, the agrifood sector appears to hold a key role in the development of a competitive advantage in the otherwise marginalised knowledge economy of the British North. In this context, university-industry research

---

[7] In 2008 25% of people aged 16 or over were obese (DEFRA 2010). According to the same statistics, obesity is expected to increase the incidence of type 2 diabetes by 70%, stroke by 30% and coronary heart disease by 20% by 2035.

collaborations appear to have a leading role in the formulation of an agrifood bioeconomy, primarily based on agri-science and agri-tech developments, which positions the region as the epicentre of the country's bioeconomy. On the other hand, agrifood inequalities are also prominent in the North of England, as a symptom of broader inequalities in the economy of the North—of which the knowledge economy constitutes a major part, and possibly, as explained above, a contributor to exacerbation of these prevailing inequalities. Such phenomena point to the importance of the North as an interesting space for further research, especially with regard to the possibilities of challenging the dominant agrifood knowledge (bio)economy vision by looking into the alternative agrifood sector, and, therefore suggesting the possibilities of an alternative knowledge economy. The rest of this chapter aims to help reclaim the knowledge economy, by specifically looking into the knowledge practices that are enacted through the initiatives involved in the alternative agro-food economy for the North of England. However, before it does so, it aims to provide further background information the specific socio-economic geography of the North West of England at the time of the rise of the knowledge economy, and of the particular alternative agri-food networks that emerged at this time and region, and therefore constitute the networks that this book focuses in order to argue for an alternative knowledge (bio)economy. The next session starts with an overview of the British alternative agro-food landscape in the aftermath of the knowledge economy agenda.

THE BRITISH ALTERNATIVE AGRIFOOD LANDSCAPE IN THE AFTERMATH OF THE KNOWLEDGE ECONOMY

In response to the above, in the early 2000s, the British agro-food landscape appeared to encompass a broad spectrum of alternative voices and strategies, which, despite their marginal role, they seemed to occupy an expanding part of the food chain, with a view to offer remedies against diverse environmental harms and economic deficits (Pretty 1995; Lang and Heasman 2004; Watts et al. 2005). UK agrifood re-localisation has been seen as providing alternatives to the high-yield productivist agro-food model. Pressures came from both British food producers and consumers. Many consumers tried to find more trustworthy sources for their food, while farmers have sought ways to add value to their produce, for example, through food promoted as local or organic, especially through direct sales. These developments have also led the government to offer grants for food producers to start anew, for example, via organic conversion (Winter 2003). The Common Agricultural Policy (CAP) decoupled single-farm payments from production levels and encouraged the use of agro-ecological methods; following the CAP reforms, the Rural Development Program England moved towards regionalisation which opened up prospects for a territorial perspective that reintegrates farming into rural development; territorially branded products, for example, locality foods, were also encouraged because of their capacity to provide added value for producers.[8]

In the context of these wider developments, a government advisory body, known as the Curry Commission, also supported such developments, by proposing economic regeneration based on reconnecting people with food production. As stated, their aim was to: '[r]econnect our farming and food industry; to reconnect farming with its market and the rest of the food chain; to reconnect the food chain with the countryside; and to reconnect consumers with what they eat and how it is produced' (PCFF 2002: 6). At the same time, the Campaign to Protect Rural England, was also supportive of local food for its potential benefits to producers, retailers and local communities. It underlined the significance of what they call as 'local food webs' for linking quality of life with quality of place, and stressed the urgency to defend these webs from the threat of supermarket chains (CPRE 2006).

Thus, local food networks were promoted as a solution for a more sustainable agro-food system, which could potentially lead to a radical break from the productivist paradigm of the agro-industrial food chain.

Alternative production methods were key in this process, with the organic and biodynamic sectors holding an important role. Organic agriculture experienced a rapid, nearly double, increase between 1997 and 2000, although still amounting to only 0.3% of the overall cultivated land (Morgan and Murdoch 2000). It was valued for its environmental qualities linked to the ideal of rebuilding healthy natural environments and soil fertility, minimising pollution, protecting and enhancing the farm environments (see Soil Association 2011). It has been acknowledged for its broader socio-environmental benefits, protecting not only natural

---

[8] Early pioneers were Southwest England and Wales (see Marsden and Smith 2005; Marsden and Sonnino 2008; Morris and Buller 2003; Ilbery and Maye 2005).

environments but also farmers from possible economic and environmental risks, as well as consumers from the health risks connected to agro-industrial production models. Biodynamic agriculture has also emerged as an alternative production method. Following a holistic approach to farming, biodynamics manifest a concern for the interrelationship between plants, animals and the soil, and claim to offer what they call 'a sound basis for sustainable food production' (see Biodynamic Association 2011). Permaculture has also been used as a more radical alternative. Considered as a 'learning-from-nature' 'ecological design process', it is based on a philosophy of 'co-operation with nature', envisioned to make the best use of available resources (see Permaculture Association 2009). It has been favoured for its distance from commercialisation, its communitarian aspects, and its environmentally friendly methods—in many cases also situating it as an alternative to organic agricultural certification processes (ibid.).

The British agrifood landscape, of course, also encompassed the wider spectrum of agrifood practices and processes that are also essential in configuring an alternative to the conventional, shorter agrifood supply chain. Especially after the incorporation and co-optation of organic food products by supermarket chains in the late 2000s—which came as a response to consumers' increasing demand for quality produce (Thanassoulis 2009; Morgan and Murdoch 2000)—the principles of re-localisation, re-spatialisation and re-connection became integral elements in configuring short food the alternative agrifood supply chains (Renting et al. 2003). Diverse agrifood practitioners came to construct such coalitions. Farmers markets, vegetable box-schemes, farm shops, market gardens, farmers and workers co-operatives, mobile groceries all constituted part of an alternative, decentralised agrifood retailing and processing system that carried the potential to rebuild physical and social proximity between producers and consumers (Feagan 2007), while minimising negative impact on both nature and the farmers.[9] At the same time, such coalitions also provided means for food producers to capture and redefine the 'market value' of their products (Smith 2006), but also for consumers to be able to meet broader socio-environmental objectives—such as accessing quality fresh

---

[9] For example, since their re-appearance in 1997, farmers markets have become a novel form of direct sales which is gradually expanding; a decade later, farmers' markets were being held at 550 locations, creating 9500 market days and 230,000 opportunities for stallholders per year and, in 2006, their total annual turnover was estimated at £220 m (FARMA 2006).

food, supporting the local economy, reducing food miles, while being reconnected with retailers and producers (Hedges and Zykes 2003).

Generally, consumers have been seen as important allies in empowering the country's the alternative agro-food sector. As part of a wider celebration of what is good in British culture, consumers have often been encouraged to buy both local and national produce. Although traditionally British food was seen as low quality, in the 2000s, this started being challenged, especially by emphasising the provenance of food ingredients. Numerous restaurants have started promoting themselves through the origin of their food. In the mass media, celebrity chefs turned into 'food activists' advocating for local seasonal produce, as well as raising concerns about animal welfare, and the environmental and economic harm from the conventional agrifood industry. At the same time, consumers have been encouraged to engage with alternative production methods themselves, by turning into what has been termed as 'prosumers'. In the late 2000s, rising food prices and environmental concerns led to an increasing demand for allotments[10] and a growth of community gardening projects. The National Lottery support has been pivotal for understanding the significance of these consumer-led initiatives in achieving broader socio-environmental and community benefits. For example, the £50 million Local Food Grants programme was established in the same decade, distributing grants to a variety of smaller food-related projects that contribute to making locally grown food accessible and affordable to local communities. The £10 million 'Making Local Food Work' was another programme that funded community and social enterprise initiatives designed to restore links between land and people—or between producers and consumers—as well as help mainstream those that are known to work (Levidow and Psarikidou 2012; Levidow et al. 2010).

Long-standing ecologically sound personal lifestyle choices have also been important in configuring the British agrifood landscape. A turn to organic and whole foods, vegetarianism and veganism, green and ethical consumerism—whose roots can also be traced back in the ecological communes and the new back-to-the-land and vegetarianism movements of the

---

[10] Allotments are predominantly inner city, municipally owned, plots of land divided into small blocks which are rented by the public and used for food production. After the Second World War these allotments had provided a real contribution to a family's diet. However, as food got cheaper and convenience food became more available, allotment use declined. Today only 300,000 remain; however, there is currently a waiting list of 100,000 people.

1960s and 1970s (see Szerszynski 1993, 2005)—has been also part of a citizen's broader political act in the quest for a transformation within the agro-food system and beyond. By turning their consumption and everyday life practices into expressions of political consumerism, citizens contributed to broadening the scope and the practices of alternative agro-food networks. Practices such as the growing of organic food in allotments and community gardens, and tree or hedge planting constituted new forms for the citizens' personal and political engagement with land and nature. Urban agriculture appeared to open up new possibilities for configuring the alternative agrifood landscape, in ways that was pivotal for not only enhancing 'consumers' appreciation and understanding of nature, but also enactment or revitalisation of different types economic relations of exchange and sharing moving beyond a money-first logic.

THE NORTHWEST OF ENGLAND: CUMBRIA AND MANCHESTER

Historically, the economy of Northwest England has been significantly shaped by industrial development. Since the industrial revolution, the city of Manchester and the surrounding county of Lancashire have had a leading role in the cotton industry and constituted the prime source of world textiles.[11] On the contrary, due to the geographical specificities of the wider Northwest region—for example, wetter climate than in some other parts of the UK and lower sunshine hours—agriculture has played a secondary role in regional economic development. According to a relevant survey conducted around the rise of the KBE economic model (NWDA 2008), in the late 2000s, the total gross margin of all farming sectors was estimated to £333 m a year. The total farmed area is 933,000 ha of which grassland and rough grazing is 784,000 ha. Horticulture and arable production remain marginal, only using 11% of the available land in the North West; dairy farming takes up 17% of the land; beef and sheep farming using the remaining land. In general, Northwest agriculture is mainly based on livestock—especially upland hill farming. Dairy farming constitutes the largest enterprise by gross margin with 67% of the total regional gross margin, whereas arable cropping in comparison has a gross margin of £43 million representing just 13% of the total (cited in NWDA 2008).

---

[11]According to Gibb (2005), by the 1830s, approximately 85% of all cotton products worldwide were manufactured in Lancashire.

This evidence illustrates the lack of agricultural diversity and food self-sufficiency of the region.

In particular, with a population of 500,000 people, the county of Cumbria is predominantly rural, mountainous and home to the Lake District National Park. Following the general geographical and agricultural characteristics of the region, upland hill farming remains the predominant feature of the agricultural landscape, whereas more income is derived from the tourism industry. While Cumbria may seem an unfavourable context for food re-localisation, after the 2001 foot-and-mouth epidemic many Cumbrian farmers were compensated for the loss of their livestock and decided to turn to production of organic or other high-quality food products. The downward price pressure from supermarkets and the threat of subsidy removal have also played a significant role in undermining the viability of Cumbrian agriculture. With an aim to gain premium prices, Cumbrian farmers established farmers' co-operatives, farm shops, farmers' markets and vegetable box-schemes, which could also pursue an empowerment of the local economy. State support has also been instrumental in the viability of their initiatives. Cumbria RDP has offered small grants for equipment to small businesses and use of renewable energy sources. Regional agencies have offered support of various kinds, which could help producer networks overcome several obstacles, such as farmers' social isolation, mutual distrust and historical dependence on CAP funds (Levidow and Psarikidou 2011). 'Distinctly Cumbrian', 'Made in Cumbria' and the Northwest Development Agency have been important support bodies, which helped to enhance the economic viability, the organisational skills and the co-operative relations among local producers, but also enhance the proximity between producers and consumers.[12] According to the Local Action Groups plan of the NWDA:

> By raising awareness of consumers of the products of the countryside, a greater loyalty is engendered and purchasing habits do change. 'Sense of place' and 'know your place' training packages have helped other areas to create local pride, develop community spirit and also contribute significantly

---

[12] For example, Made in Cumbria has provided support to four of the county's 15 monthly farmers' markets. It also organizes 'Meet the Buyer' events, helping small producers to meet larger buyers—for example, the National Trust, the Youth Hostels Association, Centre Parcs Oasis, numerous hoteliers and supermarket chains. These events help small-scale producers to gain self-confidence in dealing with buyers.

to the local tourism offer. (Cumbria Fells and Dales Local Action group 2008, p. 35)

In the city of Manchester, agriculture has traditionally played a marginal role in local economic development. Historically, Manchester grew rapidly in the nineteenth century due to the growth of the textile industry and related manufacturing. This historic economic dependence on industrialisation, as well as the wetter weather conditions and the hilly topography, has significantly slowed agricultural development in the region. Following the rise of the KBE economic model, the national statistics of the late 2000s were positioning Manchester as one of the UK's largest cities, with a population of 483,800 people. It is a metropolitan borough of Greater Manchester, was the third most populous county of England with a population of 2.6 million people. Greater Manchester also hosts a very culturally diverse population, with 66 refugee nationalities and an ethnic minority currently comprising the 8% of its total population. Manchester has not remained untouched from the KBE developments: the gradual post-war decline in industrial activity and the subsequent depopulation of city changed significantly its economic landscape. As also indicated in the council's strategic vision, the city shifted towards finance, the knowledge economy and creative industries, as key elements of its urban economic and regenerative development, despite the fact that many areas of the city were still to recover from the loss of employment in manufacturing. Agriculture still remained marginal in the county (see also Psarikidou and Szerszynski 2012).

Socio-economic inequalities, social exclusion, poor physical and mental health have been acknowledged as important factors affecting the local population's well-being. According to the Indices of Multiple Deprivation of 2011, Manchester is ranked as the second most deprived local authority in England in terms of income deprivation, third in terms of employment deprivation and fifth in terms of the extent of deprivation throughout the city (Manchester City Council 2011a). Significant inequalities and social exclusion also become apparent from indicators such as those relating to employment, education and health, such that many of the city's residents—particularly women, disabled people, black and minority ethnic communities, young and older people—are excluded from what would be considered as a reasonable quality of life (Manchester City Council 2011b). Food insecurity and poor dietary habits attributed to unequal access to food have been significant factors in boosting the above statistics.

Within the city, only 16% of adults have a balanced, healthy diet; approximately 15% of the school children are obese. Depression and other mental health problems are due to a diet low in nutritional value, whereas death rates in Manchester remain among the higher ones in the UK. As for the agro-food sector, very few residents of Manchester are employed in agriculture. At the time of the rise of the KBE, although agriculture, processing and food retail accounted for 12.5% of UK employment and 8% of the UK economy, very few Manchester residents were employed in agriculture (approximately 500 according to the 2001 Census). However, at that time, Greater Manchester also included (and still includes) some large food manufacturers, providing employment to a big number of Manchester residents. Food retail was considered as the most important part of the food chain in Manchester economy, with many residents being employed by large food manufacturing companies (Food Futures 2007). However, at the same time, there has been a growing concern about food insecurity, poverty and malnutrition, which stimulated the development of alternative agro-food strategies in the city of Manchester, also at a city council level. Environmental degradation, climate change, peak oil, and food prices were prominent citizens' concerns and therefore reasons for attitudinal and practice change, in search for more 'trustworthy' types and sources of food. State support, also in the form of a local authority strategic partnership, played a central role in coordinating and supporting the sustainable agro-food initiatives. Citizens' and grassroots initiatives, charity and not-for-profit organisations, workers' co-operatives and box-schemes have dominated the alternative agro-food sphere in the metropolitan region of Manchester and aimed to turn food into a focus in their responses to the multiple urban challenges.

Alternative Agro-Food Networks in Cumbria and Manchester

This section introduces you to two alternative agro-food networks of the Northwest England under investigation. In Cumbria, the threat to the future sustainability of the local agro-food sector has encouraged the emergence of a loose network of alternative, predominantly farmer- and retailer-led agro-food initiatives. Our research has focused on a network indicative of the diversity of alternative agri-food initiatives of a rural landscape: farmers' co-operatives, small-scale family businesses and farm shops, social enterprises, all variously utilising or supporting alternative methods of production such as organic and biodynamic, and/or alternative methods of distribution and consumption. These initiatives enact their members' aspirations for the revitalisation of the local agro-food economy, the sustainability of a rural community and the creation of a regional food culture, through the improvement of the farmers' organisational and marketing skills, the enhancement of co-operative relationships and support networks among local producers and retailers and the re-connection between local producers, retailers and consumers.

The first initiative under investigation constitutes an informal network of people—farmers, growers, processors, retailers and consumers—who have been interested in organic production in and around Cumbria. With over 70 members, it was set up in the late 1990s to provide self-help support to local farmers converting to organic production, as well as education and technical information to anyone interested in farming using organic standards. It also aimed to develop a short food-supply chain in Cumbria and build a wider food culture through public outreach, for example, via stands for local festivals, a model farm, books, leaflets, games, quizzes and farm walks (Interview with representative of organic farmers network). With similar aims, but encompassing all agricultural methods, farmers and farmer-controlled businesses in Cumbria organised themselves around another local farmer network. This second initiative under investigation was formed in the mid-2000s by a group of farmers, with an aim to maintain and develop a viable farming community across the county. Its intention was to maximise rural income via food production and other activities, promote farmer co-operation, especially in production and marketing skills, and raise consumers' awareness of local quality food through farm open days. In doing so, it wanted to raise the profile of Cumbrian farming and help farmers regain control over the food chains (Interview with representative of local farmers' network).

Around these organisations has coalesced a wider network of local—primarily producer and retailer-led—food initiatives. For the purposes of my investigation, I have focused on examples of different forms of food initiatives operating across Cumbria, while keeping diverse ways of interconnection with each other. These include: an organic farmers' co-operative in Cumbria that did direct sales through collective marketing, facilitated initial co-operative links and mutual support among local producers, while also helping them maintain a close connection with consumers providing them with a wide variety of quality, organic food and farm produce reared, grown and produced in Cumbria (Interview with representative of organic farmers' cooperative). Local organic quality food was also key for one of the first local vegetable box-scheme and farm shops in Cumbria that was also part of this study. Selling local and organic produce of theirs and other organic farmers in Cumbria and Lancashire, the aim of this initiative was to change the whole shopping experience by providing consumers with an alternative to supermarket chains (Interview with representative of farm and box scheme). The latter was also working closely with a care and social enterprise. Also set up in the mid-2000s, this initiative was focusing on growing organic produce as well as providing accredited training in horticulture. As stated by their representative, combining 'a business mission and a social aim', it was selling its produce to local farmers' businesses and farm shops, as well as providing educational visits and courses in growing, planting, harvesting and horticulture to the local community, particularly to those members of the community recovering from mental health problems. In doing so, their aim was to help them build confidence and their skills by volunteering within the business, taking on responsibilities and working in a team (Interview with representative of local care and social enterprise). This initiative was based on the land of a tenanted National Trust farm which also sold produce from the social enterprise. After many years of conventional farming, the tenants of this farm converted to organic in the early 2000s. With an aim to 'gain value' by shortening the supply chain 'in partnership with nature', this family business was committed to supporting local food producers by selling local and organically grown food. It aimed to enhance the consumers' familiarity and reconnection with local producers and local production processes, by providing educational programmes, nature walks around the farm, as well as a direct experience of farming with a tea room overlooking the milking parlour. Their farm shop was committed to selling their own milk products, as well as a range of produce from other local businesses, including flour from a local watermill (Interview with representative of farm). The watermill, also part of this study, has been one of the country's few working water-powered corn mills producing biodynamic and organic stoneground flour. Restored in the mid-1970s, the owners of the mill contributed to the conservation of the building by setting up a fully operational watermill committed to the production of high-quality flour from British grain. The initiative has been running a millshop and a tearoom, as well as educational courses and mill tours. In doing so, it aimed to build a strong local community through education and sharing, as well as build strong co-operative links among local farming businesses by not only selling, but also supplying food sources from local businesses (Interview with representative of mill).

In Manchester, poverty, unequal access to goods and services, social exclusion and health inequalities, particularly prevalent since the late 2000s, have prompted the emergence of a loose network of alternative food initiatives in various spaces across the city. The network under investigation consists of urban food producers and retailers, co-operatives and family businesses, citizen-led initiatives, charitable and non-profit organisations, which variously utilise or support alternative methods of production such as organic and permaculture, and/or alternative methods of distribution and consumption. These initiatives enacted their members' aspirations for a more socio-environmentally sustainable agro-food system with regard to climate change, peak oil, food miles, as well as broader food insecurity concerns, situated in the wider landscape of poverty and inequalities at a city level. But they also collectively performed the space of the city in different ways, through ethically motivated, embodied interactions with food, with the land, and with proximate and distant human and non-human others.

Manchester's local authority strategic partnership had a central role in coordinating and supporting the sustainable agro-food initiatives under discussion. Following its Community Strategy of 'making Manchester more sustainable' by 2015, it was providing opportunities for residents and local organisations to get involved in projects, training, activities and events around sustainable food. By bringing together health and wellbeing, local economic regeneration, food, the environment, childhood diet, vulnerable groups and transport, a wider network of local food initiatives had coalesced around them.

For the purposes of my research, I focused on a selection of these that is also representative of the diversity of initiatives: a social enterprise, initiated by the community voluntary sector, aiming to engage mental health service users, young people and the community in healthy local food permaculture growing, cooking and retailing activities and thus provide work-based learning opportunities, and 'moving-on' services which help people improve skills, confidence and overall health in order to join mainstream society (Interview with representative of social enterprise). A local partnership initiative, launched by local authority and community and voluntary organisations, for targeting and engaging the older population of Manchester in sustainable food activities (Interview with representative of local food growing projects). A small market garden providing its fresh, local produce to local businesses such as a local family-based organic non-certified box-scheme, which was also part of this study, specialising in good quality locally sourced organic produce (Interview with representatives of market garden; Interview with representatives of box scheme).

Beyond its support for an alternative supply system, in the early 2000s, the local authority partnership was also active in supporting—through funding, advice, networking and publicity—a number of community food-growing projects, many of which were directly aimed at realising social benefits. For this reason, they developed a programme that supported local food-growing projects (Interview with representative of local authority partnership). Citizen-led initiatives have been an essential part of the network: for example, the local Permaculture Network, a grassroots initiative which supported several community food-growing projects and was set up by local community members who, as described by one of our interviewees, 'fully appreciated that it [permaculture] is about understanding principles found in the natural world and ecosystems' (Interview with representative of urban agriculture network); a charitable organisation focusing on sustainable living, and a local action group focusing on sustainable neighbourhoods, both promoting sustainable living including local food and local food growing (Interview with representative of local food growing projects). A local allotments association focusing on allocating allotments to separate individuals, as opposed to the rest of the growing projects operating at a community level.

Though, how did I get to the investigation of these particular case studies? Before I move to an investigation of AAFNs as a knowledge (bio) economy, the next section aims to shed some light on the theoretical-methodological framework informing my research journey.

Researching Alternative Agro-Food Networks: Enacting Alternative Knowledge (Bio)Economies?

My research methodology is deeply inspired and driven by the theoretical underpinnings of my research study. Critical Science Studies have long argued for the incompleteness of our knowledge about the world. Haraway's 'god trick' has been key in challenging ideas of objectivity of scientific knowledge, pointing to the plurality, partiality and situatedness of our knowledge about the world in complex socio-material environments, but also underlining the performative effect of our knowledge as practices that can also enact and shape realities (Haraway 1991; Law 2004). Such ideas are not only key in shaping my research enquiry, which, in its essence aims to contribute to re-claim the knowledge economy by looking into the plurality of knowledge practices in the alternative agro-food sector. They have also been key in shaping my research methodology and methods: challenging ideas of singularity of a 'truth' out there to be discovered, realising the 'incompleteness' of a 'sound', normative scientific methodology which would lead me to the discovery of the truth, realising the complexity, or else as Law calls it the 'messiness' of the phenomenon to be unravelled, and therefore the 'messiness' of the methods that would help us better understand this world (Law 2004).

Therefore, for my research enquiry, I allowed myself to combine a diversity of qualitative, both conventional and more performative, research methods, which would help me both understand and socio-materially engage in the study of the knowledge practices involved in the alternative agro-food sector. By doing so, I was also becoming fully aware of my performative role as a researcher, who was committed to not only researching complex realities but also contributing to the enactment of certain, usually marginalised, knowledge-economic realities for the agro-food sector.

For the purposes of my research, as described in detail in the previous section, I focused on two particular 'case studies' of alternative agro-food networks, which, following 'case studies' in the social sciences, would cover a diversity of initiatives, allowing comparisons between the two as well as between 'the general' and 'the specific'; cases that may not reveal 'general', but may rework versions of the general. For this reason, I chose a rural and urban network of alternative agro-food initiatives involved in a diversity of knowledge and economic practices of production, distribution and consumption. Following ideas of 'substantive criteria' (Swanborn 2010), the location of these networks was also carefully selected, not only because of the centrality of the NW England in the configuration of the knowledge economy of England around both agriculture and food, but also for methodological purposes: the physical proximity of those case studies enabled a more uninterrupted, intimate and embodied interaction and entanglement between the researcher and the complex socio-material environments to be researched. The links appearing on these websites, as well as the references to other agro-food initiatives, gave me a first clue of the network type of relationships between these initiatives. The frequency of individuals' names appearing on the websites and the multiple appearance of the same name in different initiatives not only provided some additional evidence on the way these initiatives could constitute a network, but also helped me identify some initial contact people who could further facilitate processes of snowballing, and therefore my accessibility to relevant stakeholders. Communication via emails and phone was pivotal for introducing myself and the organisation I have been working for, as well as for making more formal arrangements with those first key people, who, as also explained below, have then directed me to other key initiatives and stakeholders that were also contacted via email and phone.

As the next step of my research, I conducted a series of interviews, starting with a small number of partly informal exploratory interviews, which were followed by a bigger number of semi-structured interviews. The first 'exploratory interviews' with key people 'in the field' were important in gathering 'reputation' and 'snowball' samples, which were useful not only in identifying more relevant actors and having access and establishing contacts in the field (Swanborn 2010, p. 46), but also for depicting the two 'AAFNs', the actual networks of initiatives and actors and the links which constituted these initiatives as a network. After these first exploratory interviews, a total of eighteen, one to two hours' duration, semi-structured interviews were undertaken in both localities, eight in Cumbria and ten in Manchester, which went beyond a set of predetermined and standardised questions in order to encourage the respondent 'to answer a question in their own terms' (May 2001, p. 121). Prior to the interviews, all interviewees were informed about the research and its purposes, and gave their consent to participate in the research. For anonymity purposes, all respondents' names have been changed and other personal or contextual information anonymised.

Following May (2001), 'semi-structured interviews' helped gather insights on the interviewees', and therefore the initiatives', 'biographies, experiences, opinions, opinions, values, aspirations, attitudes and feelings' (p. 120). Most interviews took place 'in situ', that is, at their working environment, a fact which was also important in my development of a better understanding and a personal feel of the broader socio-material environment. Some of my semi-structured interviews took the forms of 'walking interviews'. Following ideas of 'mobile methods' (Buscher et al. 2011), these interviews were usually also involving walks around 'the fields'—for example, farms, farm shops and cafes, market or community gardens. Also drawing on principles of 'sensory ethnography' (Pink 2015) and 'ethnographic interviewing' (Heyl 2001), following the interviewee has been key in developing a more situated, embedded and sensory experience and understanding of the agro-food realities at stake. In doing so, it was also envisioned to allow both me and the respondents to get equally involved in co-constructing their knowledge of the social world (Davies 1999). This particular research method also helped realise the significance of a more haptic experiential understanding and entanglement with the complex socio-material research environments under investigation.

In other words, it underlined the significance of a more ethnographic approach that would also support a more 'naturally occurring' collection of data (May 2001), and which, following the principles of 'ethnographic realism' (Hammersley 1991), would enhance knowledge of social phenomena through direct experience. In my case, this would be through my participation in a diverse spectrum of agro-food knowledge practices that are built upon multiple interactions between humans, food and nature (see also Degen 2010; Hinchliffe et al. 2005). This would also be key for the research to move towards a more-than-human methodological encounter, which, building on STS theory-methods (Law 2004), would allow my investigation of agro-food knowledge practices as socio-material practices built into complex, diverse and messy encounters between humans and non-humans, but also better understand the diverse meanings and values attributed to them. In my research, ethnographic research took the form of short-term visits to specific sites (farm-shops, co-ops, community gardens, etc.), following the principles of 'participant observation' (Lofland and Lofland 1984). During these visits, which, in some cases, were following up these interviews, I adopted different roles. For example, for my visits to the farm shops and co-operatives, I mainly obtained the role of a 'participant observer', since, following Campbell (1984), May (1991) and Watson (1994), I was not 'act[ing] as one of the group' (May 2001), but I was mainly observing the agro-food activities, by undertaking the role of the consumer; for the my visit to community gardens, I adopted the role of a 'complete participant' (see Humphreys 1970), since I attempted to fully engage in the agro-food activities by taking the role of 'consumer or citizen as producer'. This research strategy seemed to have enriched my interview data in ways that would help me argue for opening up our understanding of knowledge economy to include the greater spectrum of knowledge practices in the alternative agro-food sector, which is where our attention turns now and for the rest of this book.

REFERENCES

Bauen, A., Chambers, G., Houghton, M., Mirmolavi, B., Nair, S. Nattrass, L., Phelan, J. and Pragnell, M. (2016) Evidencing the Bioeconomy, Available online at https://bbsrc.ukri.org/documents/1607-evidencing-the-bioeconomy-report/#:~:text=The%20bioeconomy%20is%20the%20production,transformative%20processes%20using%20biological%20resources.

Biodynamic Association (2011) About BDA. Available online at http://www.biodynamic.org.uk/.

Blythman, J. (2004) *Shopped: The Shocking Power of British Supermarkets*. Fourth Estate.

Brinkley, I. (2006) *Defining the Knowledge Economy. Knowledge Economy Programme Report*. The Work Foundation. Available online at: www.flasco.edu.mx/openseminar/downloads/brinkley_S3.pdf

Buscher, M., Urry, J. and Witcger, K. (2011) *Mobile Methods*. London: Routledge Academic Publishers.

Campbell, A. (1984) *The Girls in the Gang*. Oxford: Basil Blackwell.

CPRE (2006) The Real Choice: how local foods can survive the supermarket on slaught. Campaign for Protection of Rural England. Available online at www.cpre.org.uk.

Cumbria Fells and Dales Local Action Group (2008) Development Strategy, Rural Development Programme for England, www.fellsanddales.org.uk. Available online at http://www.cumbria.gov.uk/business/rdpe/fellsanddales/fellsanddales.asp.

Davies, C., (1999) *Reflexive Ethnography*. London: Routledge.

DEFRA (2010) Food Statistics Pocket Book. Available online at www.defra.gov.uk.

Degen, M.M. (2010) The urban green: passionate involvements with urban natures. In Miles, M. and Degen, M. (eds.) *Culture and Agency: Contemporary Culture and Urban Change*. University of Plymouth Press: pp. 58–75.

FARMA. (2006) Farmers' markets in the UK: Nine years and counting. Southampton, Available online at www.farma.org.uk. Retrieved January 20, 2010.

Feagan, R. (2007) The place of food: Mapping out the 'local' in local food systems, *Progress in Human Geography* 31(1):23–42.

Food Futures. (2007) *A Food Strategy for Manchester*. Available online at http://www.foodfutures.info/site/images/stories/food%20futures%20strategy%202007.pdf.

Gibb, R (2005) *Greater Manchester: A panorama of people and places in Manchester and its surrounding towns*. Myriad, 13.

Hammersley, M. (1991) Some reflections on ethnography and validity, *Qualitative Studies in Education* 5(3):195–203.

Haraway, D. (1991). 'Situated Knowledges: The science in feminism and the privilege of partial perspective In D. Haraway (Ed.) *Simians, Cyborgs and Women: the reinvention of nature*. New York: Routledge.

Hedges, A. and Zykes, W. (2003) *Local food: a report on qualitative research*. London: Food Standards Agency. Available online at http://www.food.gov.uk/multimedia/pdfs/localqualitative.pdf. Retrieved on January 20, 2011.

Heyl, B. (2001) Ethnographic Interviewing. In P. Atkinson, S. Coffey, and Delamont, S. (Eds.) *Handbook of Ethnography*. SAGE: pp. 369–383.

Hinchliffe, S., Degen, M., Kearnes, M. and Whatmore, S. (2005) Urban wild things: a cosmopolitical experiment, *Environment and Planning D: Society and Space* 23 (5): 643–658.

Humphreys, L. (1970) *Tea Room Trade*. London: Duckworth.

Huxley, R. (2003) *A Review of the UK Food Market*. Report for Cornwall Agricultural Council and Taste of West. Available online at http://www.objectiveone.com/ob1/pdfs/uk_food_market_review.pdf.

Ilbery, B. and Maye, D. (2005) Alternative (shorter) food supply chains and specialist livestock products in Scottish-English borders, *Environment and Planning A* 37(4):823–844.

Lang, T. and Heasman, M. (2004) *Food Wars: Public health and the battle for mouths minds and markets*. London: Earthscan.

Lang, T., Barling, D. and Caraher, M. (2009) *Food Policy: Integrating Health, Environment and Society*. Oxford: Oxford University Press.

Law, J. 2004. STS a Method. Lancaster University. Available online at http://heterogeneities.net/publications/Law2015STSAsMethod.pdf.

Levidow, L. and Psarikidou, K. (2011) Food Relocalisation for Environmental Sustainability in Cumbria, *Sustainability* 2(1):692–719.

Levidow, L. and Psarikidou, K. (2012) Making Local Food Sustainable in Manchester. In Viljoen, A. and Wiskerke, J.S.C. (eds.) *Sustainable Food Planning: Evolving Theory and Practice*. Wageningen: Wageningen Academic Publishers: pp. 207–220.

Levidow, L., Price, B., Psarikidou, K., Szerszynski, B. and Wallace, H. (2010) Urban Agriculture as Community Engagement, *Urban Agriculture Magazine* 24:43–45.

Lofland, J. and Lofland, L. (1984) *Analysing Social Settings: A Guide to Qualitative Observation and Analysis*, 2nd edn. Belmont, CA: Wadsworth.

Manchester City Council. (2011a) *Indices of Multiple Deprivation 2010. Analysis for Manchester*. Available online at http://www.manchester.gov.uk/downloads/download/414/research_and_intelligence_population_publications_deprivation.

Manchester City Council. (2011b) *Manchester: A City for Everyone. Promoting Equality and Inclusion*. http://www.manchester.gov.uk/download/210/a_city_for_everyone.

Marsden, T.K. and Smith, E. (2005) Ecological entrepreneurship: sustainable development in local communities through quality food production and local branding, *Geoforum* 36:440–451.
Marsden, T.K. and Sonnino, R. (2008) Rural development and the regional state: Denying multifunctional agriculture in the UK, *Journal of Rural Studies* 24:422–431.
May, T. (1991) *Probation: Politics, Policy and Practice.* Buckingham: Open University Press.
May, T. (2001) *Social Research: Issues, methods and process.* Buckingham, Philadelphia: Open University Press.
Mintel (2001) *Regional Eating and Drinking Habits.* Market Intelligence. December 2001. London: Mintel International Ltd.
Morgan, K. and Murdoch, J. (2000) Organic versus Conventional Agriculture: Knowledge, Power and Innovation in the Food Chain, *Geoforum* 31:159–173.
Morris, C. and Buller, H. (2003) The local food sector: A preliminary assessment of its form and impact in Gloucestershire, *British Food Journal* 105(8):559–566.
NWDA (2008) More resources to grow rural economy, *315 Magazine*, Issue 15, June, 2008. Available online at http://www.nwda.co.uk/news%2D%2Devents/features/building-communities/resources-grow-rural-economy.aspx. Retrieved on July 20, 2008.
OECD (2012) Gross Value Added. Glossary of Statistical Terms. OECD. Available Online at http://stats.oecd.org/glossary/detail.asp?ID=1184. Retrieved on 25 March 2012.
PCFF (2002) *Farming and Food: A Sustainable Future*, Policy Commission on Farming and Food, chaired by Sir Donald Curry, London: Cabinet Office.
Permaculture Association (2009) Knowledge base. Available online at http://www.permaculture.org.uk/.
Pink, S. (2015) *Doing Sensory Ethnography.* California: Sage Publications.
Pretty, J. (1995) Participatory learning for sustainable agriculture. *World Development* 23(8): 1247–63.
Psarikidou, K. and Szerszynski, B. (2012) Growing the Social: Alternative agro-food networks and social sustainability in the urban ethical foodscape, *Sustainability: Science, Practice and Policy* 8(1):30–39.
Renting, H., Marsden, T. and Banks, J. (2003) Understanding alternative food networks: exploring the role of short food supply chains in rural development, *Environment and Planning A* 35(3): 393–411.
Smith, A. (2006) Green niches in sustainable development: the case of organic food in the United Kingdom, *Environment and Planning C: Government and Policy* 24: 439–458.
Soil Association (2011) *What we do.* Available online at http://www.soilassociation.org/.
Swanborn, P. (2010) *Case Study Research: what, why and how?* London: Sage.

Szerszynski, B. (1993) Uncommon Ground: Moral Discourse, Foundationalism and the Environment Movement. Unpublished PhD Thesis, Lancaster University.

Szerszynski, B. (2005) *Nature, Technology and the Sacred*. Oxford: Blackwell.

Thanassoulis, J. (2009) *Supermarket Profitability Investigation*. University of Oxford, January 2009.

Ward, N. (1994) *Farming on the treadmill: agricultural change and pesticide pollution*. Unpublished PhD Thesis, University College London.

Watson, T.J. (1994) *In Search of management: Culture, Chaos and Control in managerial World*. London: Routledge.

Watts, D.C.H., Ilbery, B. and Maye, D. (2005) Making reconnections in agro-food geography: alternative systems of food provision, *Progress in Human Geography* 29(1): 22–40.

Winter, M. (2003) The policy impact of the foot and mouth epidemic, *Political Quarterly* 74(1): 47–56.

CHAPTER 5

Re-Thinking the Knowledge Economy Through Alternative Agro-Food Networks

This fifth chapter conceptualises alternative agro-food networks as an alternative knowledge economy. Drawing on fieldwork with urban and rural agro-food initiatives in the UK, it presents the ways scientific and local, explicit and tacit forms of knowledge get conceptualised, perceived and performed through engaged actors' discourses and practices. As already discussed, alternative agro-food networks are configured as alternatives to the conventional agro-food system. Claimed to introduce a radical discontinuity with the technological innovation of productivist agro-food knowledge networks, they were considered as carriers of a 'radical innovation' (Dosi 1988; Johnson 1992), which could open up pathways for a different learning process where the acquisition of new knowledge could stem from the 'forgetting' and 'creative destruction' of old knowledges, as well as the resurrection of the local context-dependent knowledges.

This chapter aims to shed light on the alternative innovation potential of alternative agro-food networks, by specifically focusing on the different knowledge practices that shape and configure the alternative agro-food sector of the Northwest of England. In order to do so, it is mainly divided into three distinct empirical sections that provide detailed analysis of the knowledge patterns and production processes within the alternative agro-food sites of production, distribution and consumption. The first section focuses on the site of agricultural production and revolves around the investigation of the complex ways scientific knowledge is both perceived and practiced by AAFN stakeholders. Focusing on distribution and consumption, the other two sections re-iterate this point, by providing detailed evidence of the fruitful dialogue, recombination and adaptation of different forms of knowledge within the alternative agro-food sector. This leads us to the final section of this chapter, which discusses the ways alternative agro-food networks constitute a knowledge economy. It identifies the characteristics through which an alternative agro-food economy resembles a knowledge economy, but also the ways it is different from its mainstream capitalocentric understanding: an analysis that also sets the grounds for understanding the alternative agro-food economy as a knowledge economy, but also for reclaiming the knowledge economy in the final chapter of this book.

Knowledge Production in the Alternative Agro-Food Networks of Northwest England

Science and the Alternative Agro-Food Sector

Scientific knowledge can be key in understanding processes of knowledge production within the alternative agro-food sector. As already discussed, science as the dominant, objective and universal form of knowledge has been key in framing discourses of agri-food innovation, and therefore establishing certain knowledge hierarchies within formulations of the knowledge (bio)economy. Therefore, exploring understandings, experiences and perceptions of scientific knowledge within the alternative agro-food sector appears key for better understanding the different knowledges and the ways they come together and relate to each other for the production of knowledge for agriculture and food in the so-called alternative sector.

As a first reaction to my question, agro-food practitioners situated their perceptions of scientific knowledge in the broader political economy of research and innovation for agriculture and food. For most of them, 'science' appeared to be mainly associated with the advanced techno-scientific developments dominating the conventional, agro-industrial food system. They underlined the role of corporate powerful stakeholders in dictating a techno-solutionist approach to the role scientific knowledge for agriculture and food, and therefore discussed the homogenising effect of scientific knowledge for agrifood knowledge systems, its role and accountability for the gradually disappearing traditional farming skills. As described by two practitioners involved in urban agricultural schemes in Manchester illustrated:

> After the industrial revolution, we created these problems; technology is like a genie in the box ... think how much people have lost, they so much rely on technology. The solution is always a technological solution. (Interview with representative of local allotment association)

> research is connected to commercial interests. Conflicts of interests are the funders and what the resource says. Research on sustainability has been funded by Tesco, and, thus, compromised from the start. There should be room for good, reliable information. (Interview with representative of market garden)

Therefore, science has been very much perceived as a highly rationalised, universalisable form of knowledge driving techno-scientific innovation for agriculture and food, under the principles of productivity and growth. It was conceived as key driver for processes of commodification within agriculture and food, in which, similarly to the knowledge economic vision, scientific knowledge not only contributes to such processes, but also constitutes one of the commodities to be sold for profit. However, as also hinted above, practitioners also suggested challenging the domination of science by a narrow techno-scientific, productivity-driven, profit-maximising logic. They approached scientific knowledge-making as a value-laden process, driven by a diversity of motivations and interests. In this context, they challenged ideas of 'scientific objectivity' and impartiality attributed to scientific knowledge (see Haraway 1991), but also pointed our attention to the binaries that exist within science-making processes. This dual role of scientific knowledge has also been depicted in the discourse of one of the community food activists involved in the community garden initiatives in Manchester:

> science isn't a bad thing, is it? It's like anything that contributes to knowledge and understanding has got to be a good thing. Then you have got like the scientific industry—it's kind of like science ... like someone says to you 'here is a scientific project, we need someone to research this, off you go', and it's like we are developing nuclear technology and yet the result is going to be a bomb which is going to kill half a million people. (Interview with representative of urban agriculture network)

Here, science is conceived as a value and motivation-laden project, highly dependent on its users, applications and outcomes. Following Michael Polanyi (1967), for most agro-food practitioners, science has been configured as the outcome of a personal act that is highly charged with scientists' personal feelings, commitments, passions and responsibilities for the pursuit of what could be conceived as a hidden, value-free truth. Phrases such as 'sound science' (Interview with representative of organic farmers network), 'best science' (Interview with representative of local care and social enterprise) and 'good science' (Interview with representative of market garden) used by agro-food practitioners also reflect multiple understandings and aspirations for the role that science can play in configuring an alternative knowledge-based agro-food economy. The example of organic agriculture has been key in depicting practitioners' binary understandings of science, as well as attempts for re-defining the value of science (see also Wilsdon et al. 2005):

> Organic research is useful thing. It'd be good. It's good science, multidisciplinary based on empirical material and not on a political agenda. (Interview with representative of market garden)

Therefore, for many alternative agro-food practitioners, realising the politics of science is key for moving towards an alternative science-based model for agriculture and food, in which, as will be further discussed below, scientific knowledge becomes situated in a much more complex socio-material landscape, co-habited by different forms of knowledge. In this context, as stated above, de-contextualising scientific knowledge from certain political economic interests appears pivotal for reconfiguring a differently (re-)politicised scientific knowledge.

Knowledge Production at the Agro-Food Production Site: Science and Beyond

My exploration begins with the agro-food site of production, with a view to investigate the role of science as well as other knowledges in configuring the alternative agro-food economies as knowledge economies. It specifically draws on findings from alternative agro-food networks in Cumbria and Manchester,[1] focuses on diverse farming practices—from organic to biodynamic and permaculture—and unpacks the complex knowledge production processes involved in the alternative agro-food sector.

Organic farming has been a key example variously used by agro-food practitioners for opening up to a more multi-faceted understanding of science. Soil Association, the UK's leading organic charity organisation, has been pivotal at underlining the 'scientific' nature of organic farming knowledge practices. While distancing from the science involved in the conventional agricultural methods, its description of organic farming practices also reveals the centrality of science in organic farming, in terms of soil, biodiversity preservation, crop cultivation and growing techniques. As said:

> organic farming practices are those which prohibit the use of conventional agricultural methods—such as use of pesticides, artificial chemical fertilisers, drugs, antibiotics, as well as genetic engineering technologies. Instead, they are based on the *cultivation of a nutrient-rich and fertile soil* and the *preservation of wildlife and animal welfare*, through the adoption of alternative growing, cultivating and preventative techniques—such as *rotating a mixture of crops* using *clover to fix nitrogen from the atmosphere, moving animals to fresh pasture* and keeping smaller herd size, and promoting free-range life for farm animals. (Soil Association 2011a—italics added for emphasis)

In Cumbria, many practitioners were involved in organic farming. Jane was one of the interviewees working as a co-ordinator for a local organic farmers network. She is an agricultural scientist who has family bonds with rural farming practices. While she was studying agricultural science and working for a project in the Middle East, she re-directed her interest towards organic agriculture as a more eco-efficient way of cultivation widely independent from the use of agrochemicals and drugs for animals. Building on her scientific background, for her, organic farming is a science-based model, however, differentiated from the science of conventional farming methods, primarily in terms of the values and principles represented within it, which is what also makes her talk of the 'sound science' involved in organic knowledge practices. As she claimed:

---

[1] Please note that all names of research participants have been changed for anonymity purposes.

> [organic farming] is more science-based in the fundamentals. Conventional uses more science in what is brought in to the farm, but in organic the fundamental system is based on sound science. (Interview representative of organic farmers network)

John, the chairman of the local farmers' support network in Cumbria, who is himself working as a conventional hill-farmer, indicated the ways that organic agriculture could even be perceived as having a greater scientific basis than conventional farming methods. In our visit to his farm, while showing to us his animals, his farming tools and machinery, he explained his scientific understanding of organic hill-farming. Such discussions have also been key in understanding agricultural science as an embodied way of knowing that is intrinsically connected to everyday farming practices and socio-material engagements with nature. As said:

> organic farmers have a better understanding of grass land management certainly, they apply certain scientific techniques to make grass grow ... they come from a more scientific standpoint than normal farmers do. (Interview with representative of local farmers' network)

Both above observations encourage us to consider the centrality of scientific knowledge in the configuration of organic farming as an agricultural knowledge system. However, they are also important in understanding the inherent dualisms within science, with organic science constituting a 'science-in-the making' knowledge system (Latour 1993), in which science is done in practice, as part of much more complex interactions, hybrid and messy socio-material entanglements involving humans and non-humans. Tom, the co-ordinator and manager of a local care and social enterprise project, has complemented such understandings, by pointing to the hybrid nature of organic farming, as a knowledge system that is both based on what can be considered as 'agricultural science' but also acknowledging the embeddedness of this scientific knowledge in everyday farming practices. According to his own understanding of the relationship between organic farming and science:

> With the organic model there is a scientific rationale for growing in that way and there are techniques and the thinking behind of how you rotate from one crop to another crop to another, is based on science, so in that respect it's a scientific process. (Interview with representative of local care and social enterprise)

Organic standards[2] have also been used as an example for understanding the scientific nature of organic agriculture, but also the embodiment of such processes in what was described as 'common sense' knowledge and practices. In this context, organic farmers have also been configured as 'scientists'—in ways that also contributes to challenging ideas of experts in agricultural science, and considering the broader spectrum of experts and knowledges that become significant in the making of science for agricultural practices. As discussed by two of the research participants from Cumbria:

> The standards are there to ensure we do things right ... I don't have an issue with farming to those standards. The science behind those standards seems to be common sense. (Interview with representative of local farm and box-scheme)

> Organic is about designing closed systems ... You must be a good scientist to understand the systems. If you don't, it's very difficult to manage an organic farm. (Interview with representative of organic farmers' network)

The above descriptions are key for opening up our understandings of scientific knowledge, as a more hybrid, complex form of knowledge that is embedded in tacit knowledge, the practical experiences and skills acquired by farmers during their personal, constant interaction with land, or as two interviewees mention, their close 'work with nature' (Interview with representative of local farmers' network) and 'understanding of the soil' (Interview with representative of market garden). Such understandings are important for configuring scientific knowledge as a certain stage in the

---

[2] Organic standards are the rules and regulations that define the ways an organic product must be made. The EC regulations 834/2007 on organic production and labelling of organic products, as well as the supplementary EC regulations 889/2008, 1235/2008 and 710/2009 have been set to lay down the implementing rules for the 834/2007 regulation. According to Soil Association—one of the key organic certification bodies in the UK—organic standards are set to cover all aspects of food production, from animal welfare to wildlife conservation, to food processing, to food packaging. Following the EC regulations, the Soil Association employs committees who are responsible for the technical evaluation of standards in specific areas. After sending some inspectors to each site, the inspector writes a report that sets the standards each enterprise needs to comply with in order to be certified. Once the standards are met, the enterprises need to pay an annual certification fee in order to be licensed by the Soil Association. An annual inspection takes place for the continuous meeting of standards (see Soil Association 2011b, 2011c).

much more complex and diverse process of knowledge production, which is based in a combination and sharing between a broader spectrum of knowledges already embedded and embodied in farmers' daily practices. In doing so, the science, also related to the codification and standardisation processes, challenges existing knowledge hierarchies between explicit and tacit forms of knowledges. In this particular case, standardisation processes are not only understood as embedded into farmers' tacit knowledge practices, but are also configured as a way of translating, communicating and replicating such complex knowledge intersections amongst wider farming communities of practice. In this context, organic science appears to constitute a 'knowing process' where 'know why', 'know what' and 'know how' seem to interchange and co-exist; it can be seen as a new 'cognitive space', where knowledge is not only local, but also 'located', 'situated' and 'situating'[3]—going beyond the physical boundaries of a locality and creating a 'knowledge space' through the construction of an assemblage of linked sites, people and activities (see Turnbull 1997).

Practical experience, embodied and embedded forms of knowledge appeared key in most Cumbrian farmers' descriptions of their farming knowledge practices. Despite the acknowledged centrality of scientific knowledge in alternative production practices, phrases such as 'traditional' or 'old' revealed the sense of a 'continuity' between traditional and new forms of knowledge, re-invigorating agricultural knowledge production an ongoing, old and new, experience-based learning process (Fonte and Grando 2006). Many practitioners in Cumbria underlined the historical roots of their knowledge practices, their connection to the past experiences and knowledges of older generations, primarily linked to pre-agro-industrial agricultural practices. In this way, the alternative agro-food production system can be seen as a new hybrid knowledge system that is based on the transfer, exchange and sharing between old knowledges and practices, situated personal experiences and scientific, codified knowledges:

> Well it's old, like my grandfather had to use these methods ... but I'm using my scientific knowledge as well ... I'm always asking my dad: How did you do this before you had chemicals? (Interview with representative of organic farmers network)

---

[3] See p. 78 of Chapter two of this thesis.

> it is something old, like farming in the time of my father ... it brings me closer to nature and the land ... [it] is like a nature reserve. (Interview with representative of organic farmers' cooperative)

> the Mill essentially is very traditional ... people mill with stones all over the world for hundreds of years as we do here ... what we wanted to do was to have tradition but also move on and be part of the present. (Interview with representative of mill)

Through such understandings, traditional knowledge is directly connected to the experiences and knowledges of older generations, tacitly embedded in farming practices of pre-industrialised agro-food production processes. In this way, alternative agricultural practices are configured as an alternative, 'post-industrial', science-based production system that is based on the recovery of past as well as present on-the-ground farming experiences, knowledges and skills. It is a new knowledge-based farming model, which is based on a fruitful combination and dialogue between a diversity of experts, and a diversity of explicit and tacit forms of knowledge, in which explicit forms of knowledge become rooted in experiential on-the ground knowledge practices, but also become tacitly manifested in organic farmers everyday practices. As Tom from the local social enterprise described:

> it's much more based in practical experience, that practical experience is based in scientific knowledge but I think people who are involved don't need to know the science. They need to know how to grow carrot or cabbages, how to deal with a particular pest, but I don't picture science being involved. (Interview with representative of local care and social enterprise)

Personal practice-based experiences have been configured as important for a lot of Cumbrian farmers, describing farming as an ongoing 'learning by doing' process, based on a day-to-day experiential understanding of the local environment and a more direct connection with the land and soil. Such experiential, situated understandings of farming knowledge have also been key for those farmers who admitted to not have any prior farming experience. Chris and Mary are a couple of farmers running a farm-based box-scheme. They used to be teachers, working and bringing up their family in a big city. They moved to Cumbria in the mid-1990s and resorted to farming, as a way of pursuing an alternative lifestyle that would reconnect them with land and food, and expressing their care for the environmental and health impact of conventional agrifood practices. Emily, Richard and Helen had similar stories to tell. Emily, member of a local farmers' co-operative, used to also live in London before moving to Cumbria to change lifestyle by growing food on her husband's family land. Richard and Helen moved also from London in the mid-1970s when they decided to 'save' a watermill which was under a plan of future demolition. For all of them, farming has been described as an ongoing learning process, based on continuous, personal, experiential interaction with nature, food, animals and soil.

In Manchester, farming knowledge has also been configured as practiced and embodied, situated and embedded in a more complex and ongoing socio-material entanglement between humans and non-humans. Organic agriculture and permaculture were the two main farming systems employed by Manchester's agro-food initiatives. Brian and Clare are a couple of urban farmers running a market garden in the broader Manchester area. After negotiating with the local council, they turned a derelict site in their urban garden project. When I visited them, Brian walked me around their farm, showing to me his well-looked-after newly planted lettuces, while sharing some basic tips on each of his plants' needs. He talked to me about on-site composting, crop rotation, green manure crops, as well as some more specialised techniques—such as horticultural fleece,[4] mesh cover, polytunnels and crop irrigation—all being important methods for maintaining soil fertility and plant health. For Brian, working in the field has been a constant encounter with challenges: they had to get to know their land, the soil and the local climate needs. Seeking advice and learning from other local farmers has also been important. As said, 'understanding of the soil' was an ongoing processes based on both personal engagement with land as well as other people, particularly those farmers with a longer-term experience and knowledge of the local environment and nature. In this context, a reconfiguration of understandings of 'experts' also prevails to include a broader spectrum of agents and ways of

---

[4] Horticultural fleece is a thin fabric used to cover early crops and other delicate plants in order to raise temperature in a variety of settings and protect them from cold weather, frost and insect pesticides. While allowing the penetration of air and rain it creates a microclimate around the developing plants, allowing them to grow faster than they do when unprotected (Royal Horticultural Society 2011).

knowing that is based on a synthesis of plural knowledges and skills. As he said:

> We have been gathering knowledge and patching knowledge from people who already knew. Organic is more complicated than conventional. There's a high level of knowledge, whereas in conventional, there are specialists with special tasks. Organic is based on smaller scale and aims to understand more variables ... It is based on synthesising knowledge, and the soil has a critical difference ... Soil is the basis, understanding the soil. (Interview with representative of market garden)

Coming from a similar standpoint, 'understanding how nature works' has also been key in conceptualising permaculture. Originally conceived in the mid-1970s as 'permanent agriculture', permaculture represents an approach for designing human settlements and agricultural systems 'in harmony with nature' (Permaculture Association 2011a, b). It is configured as an agricultural knowledge system that is based on a constant observation of and engagement with the natural world, embedded in the ecological processes of plants, animals, their nutrient cycles (ibid.). According to the Permaculture Association description:

> permaculture is about creating sustainable human habitats by following nature's patterns. It uses the diversity, stability and resilience of natural ecosystems to provide a framework and guidance for people to develop their own sustainable solutions to the problems facing their world, on a local, national or global scale. It is based on the philosophy of co-operation with nature and caring for the earth and its people. (Permaculture Association 2009)

Manchester has been hosting a number of social enterprises and citizen-led food growing initiatives based on the principles of permaculture. Organised around Manchester's local permaculture network, such projects were not only intended to 'grow' people's knowledge around nature and food, but also healthy communities (see also Psarikidou and Szerszynski 2012). I was excited to join one of the local food growing project's planting day. The project had already started earlier in the year, when the first varieties of trees were planted. Following the principles of permaculture forest gardening—efficiency of space, multi-layering, mimicking a woodland structure, using edible plants—our task was to help cultivate and transplant four quadrants of the forest garden. People from different ages, ethnicities, backgrounds, and gardening experiences came together to enrich the vegetation of the garden. Once we got there, we all became aware of the work awaiting us: diverse tools—mainly spades, hoes—as well as clothing for the not well-prepared—such as gloves and wellington boots. The more experienced members of the group were there to advise, offer help and instructions. This provided the foundations of a much more complex knowing processes, based on complex web of interactions between a diversity of expert and lay permaculture gardeners, as well as people, plants, land and soil.

The 'permaculture design science' appeared central to the work that the majority of the amateur gardeners needed to do. William, one of the local network members, made clear to me the importance of spacing between the trees, particularly designed so that other species and smaller plants, herbs and trees can also get planted. This helped configure permaculture as an alternative 'science model' based on a very carefully specified design of natural ecosystems. However, his description also revealed the significance of situated knowledges, grounded on a more embodied experience of local environments, their specificities and needs. As explained to me, the spot for planting each plant was very much carefully selected, based on a good knowledge of their special needs for sun, water and soil, as well as a careful observation of the plots and identification of those specific parts of it that would cater for the plants' needs (e.g. the natural water flow, the hours and angle of sunshine at that particular site). As we kept on working, permaculture was becoming clearer to my mind as a 'science-in-the-making' model (Latour 1993), rooted in personal experiences and situated knowledges, as well as ongoing observations and interactions with a particular locale, its natural specificities and needs. As described to me:

> Well the thing with permaculture is it's sometimes described as a design science. So it's about whole systems thinking and it's um you can apply the science to a small area of land or you can apply it to a regional scale if you had the resources and the mind to do that ... these are all elements of the system so it's how you arrange the different elements and link them together in beneficial ways so that you are reducing the amount of effort you have to put in as a human being interactive in this system and also at the same time trying to increase the amount of yield and diverse yields you get from the system. (Interview with representative of urban agriculture network)

Therefore permaculture has helped open up to a different, more contextual understanding of science that could build upon personal interactions, observations and experimentations with the land and nature. This was becoming clearer in my mind while I was trying to help with planting. Most people involved were amateurs, who, as I was told, had no hands-on experience in gardening. The most experienced of them, members and friends of the network, described their food growing practices as part of an ongoing learning process, in which agricultural knowledge was gradually embodied in them through their personal interaction with land and experimentation with nature. Many people confessed that they didn't know the permaculture principles but they still knew what to do and how to do it. 'Working with nature ... not man controlling nature' has been a moto used by people, underling the significance of permaculture as an agrifood knowledge system in harmony with nature.

The significance of practical knowledges and skills was also stressed by Laura, another member of the local urban agriculture network. In her understanding, food is a process of knowledge sharing, not only among members and friends of the network, but also between different generations, particularly younger and older. In this context, permaculture has also ben configured as an intergenerational knowledge system, based on the combination of old and new knowledges, and the revitalisation of those skills which may have been lost during what Giddens calls (1991) modernity's de-skilling of everyday life. As she said:

> it seems ridiculous to me that I am running workshops about food growing when I have only been doing it 3 years. It's like the blind leading the blind almost, whereas you have this whole load of people that have been doing it for years and got a lot of knowledge but there is no kind of forum for them to be able to teach people about it ... I was talking to [name of food growing project member] about developing some sort of project that was deliberately about getting the knowledge from the older guys from the allotments to teach the younger ones that are coming along, because potentially I think that would help smooth the social vibes and also would actually pass on some of the knowledge. (Interview with representative of local food growing projects)

In this way, permaculture reinforces the idea of knowledge as practice that is embedded in a broader context of socio-material entanglements and a closer connection with past experiences, habits and traditions.

Overall, we could claim that in both cases farming in rural Cumbria and urban Manchester, different knowledges and skills co-exist and contribute to the transformation of production spaces into 'cognitive spaces' for agricultural practice. Explicit and tacit forms of knowledge—scientific and lay, new and old, modern or traditional, local and situated, practical experiences and 'learning by doing'—seem to play an equally important role in configuring the alternative agro-food production spaces as knowing spaces. In this context, science is configured as another form of knowledge, which is in constant communication with those usually 'othered' or marginalised forms of knowledge. It goes beyond its narrow understanding as the dominant, complete and codified form of knowledge, usually affiliated with mainstream techno-science driven innovation trajectories for agriculture and food. It is the form of knowledge that, also following Polanyi's approach, is tacitly embedded in a broader spectrum of practices and experiences, generated through complex interactions, observations and work with nature. It resembles Latour's 'science-in-the-making' that is as the outcome of a democratic process where a greater diversity of knowledge agents can be gainfully and authentically involved in the production of scientific knowledge.

Following Epstein's idea for more inclusive bottom-up science (2007), as well as Callon et al.'s (2009) 'hybrid forums', we could understand the alternative agrifood spaces as the knowledge spaces that go beyond the traditional dualisms between science and society, nature and culture. They constitute the knowledge spaces in which agrifood knowledge is produced and practiced as part of a much more complex process of deliberation between a plural and diverse set of knowledges.

Knowledge Production at the Distribution Site

In Cumbria, farmers' isolation has been a significant factor affecting as well as shaping local processes of agrifood knowledge making and sharing. Due to Cumbria's hilly topography, farmers' interaction and internal communication have been limited, therefore further affecting the possibilities for collaboration, collective support and learning. In this context, the local farmers' networks have been key for enabling such channels of communication as well as enhancing farmers' marketing, collaborative and communicative skills. 'Break[ing] down the isolation' (Interview with representative of local farmers' network) and 'improv[ing] communication' (Interview with representative of local organic farmers' network) have been key objectives of Jane's and John's work, who were representatives of these two networks. As John described:

> it is an alternative to individual working that's what it is, it's an opportunity to be part of an organisation that's trying to make your life better ... it's certainly made people more aware of the marketing opportunities that are out there ... encourage people to be more cooperative in the way they work, not just in marketing but in production as well ... The members of our network have become a lot more aware of the benefits of working cooperatively and really taken control of their own futures. (Interview with representative of local farmers network)

Helen, one of the first organic farmers in Cumbria, had underlined the importance of local organic farmers' network in empowering local organic farmers' marketing skills. As she explained, gradually, the organic farmers succeeded in breaking their isolation, getting to know and communicate with each other. They developed their own self-help support network, by having regular meetings, sharing knowledge and ideas, buying and supplying produce and main ingredients from each other—for example flour and wheat from local farmers, milk from the local farm shop and vegetables from the social enterprise under investigation (Interview with representative of mill). As Emily, member of local organic farmers' co-operative, said, such processes of 'working together' have also been key for enhancing farmers' knowledges and skills, about both production and distribution practices. As claimed:

> It takes time to get to know people and build up relationships ... It is important to work together. The other day I wouldn't have got my hay made without being able to ask fellow co-op members to help me. (Interview with representative of organic farmers' cooperative)

Therefore, farmers' 'know who', in other words their social skills and interactions with each other, become important in enhancing their 'know how', also by gaining access to the 'know how' of the others. In this context, agrifood knowledge production is configured as part of a much more complex, ongoing process of learning embedded in relations of communication and knowledge sharing between a diversity of actors who come to configure their own 'communities of practice' (Blackler 1995; Brown and Duguid 1991; Lam 2000).

However, the enhancement of the farmers' social skills has been important not only for enhancing co-operative links, but also fostering socio-spatial proximity with consumers. Here the development of alternative marketing skills also had a key role to play. As Chris and Mary from a local farm-based box scheme explained, in the early days of their box-scheme, the supermarket digitised marketing and retailing platforms provided good grounds for configuring their own systems of home delivery as well as direct selling. These have been pivotal for identifying and developing new ways of re-connecting with consumers through the use of online platforms and their role in enhancing the know who, know how and know where of both producers and consumers. As one of them said,

> people come to the shop and see what we actually do and as a result purchase stuff ... or box customers then come and visit the shop. (Interview with representative of local farm and box-scheme)

Therefore, online channels of communication have been important in not only enhancing the collaborative and networking links between local agrifood practitioners, but also for connecting more directly and broadly with consumers. In this context, agrifood practitioners' as well as consumers' digital skills have been important in supporting this alternative agrifood relocalisation system by creating new markets for those farmers while reconnecting local farmers, retailers and consumers. Here, it is also interesting to note the role of the supermarket ICT-based marketing and retailing system in up-skilling the consumers, and therefore opening up new possibilities for the ICT-based innovative practices for the alternative agrofood initiatives. As Mary explained:

> In some ways Tesco's home delivery has benefited us in that it has introduced a new way of shopping to people ... We have spent ages marketing ourselves as a box scheme and marketing is expensive ... Now, 50% of our customers are from the online shop. (Interview with local farm and box scheme)

In this context, the alternative agrifood distribution system is configured as a hybrid knowledge system, based on a combination of diverse off-farm knowledge practices including ICT skills. This also affirms those networks' complex relationship to science and technology, manifested in the broader spectrum of technological innovations and scientific knowledges, sometimes also broadly adopted by the conventional food sector.

However, producers' and retailers' educational and communicative skills have also been important for pursuing a greater socio-spatial proximity between them and local communities. The networks of farmers in Cumbria have been instrumental in organising seminars, educational courses, public awareness campaigns, farm walks and visits envisioned to enhance people's knowledge around food. The latter has become explicit in a number of food initiatives around Cumbria. One of the local farm shops and tea rooms has been organising diverse educational activities, such as farm visits and farm walks, educational courses and seminars, as part of their collaboration with a local social enterprise particularly working with mental health service users and other vulnerable members of the local community.

Emily from a local organic farmers' cooperative has also been holding open farm visit days and educational courses for students from deprived schools across Britain. When I visited her, Georgia, a schoolgirl from London, was also there staying and working at the farm for some weeks over the summer holiday break. Emily invited us to her kitchen table, where we started our discussion, while she was transferring redcurrant marmalade and crab apple jelly from some big pots to marmalade jars ready to be sold. While I was becoming witness to her excellent cooking skills—here it is also worth mentioning the Emily is primarily a livestock farmer—she explained to me the way organisational skills had become important part of her everyday farming practices. At some point, she went to her office and brought big cardboard boxes with papers. As she explained, organic and educational accreditations, animal slaughtering arrangements were gradually transforming her into a clerical worker (Interview with representative of organic farmers' co-operative), or what Davis and Hinshaw (1957) had described as a farmer in a 'business suit'.

In Manchester, a more diverse spectrum of managerial, logistical, marketing and bureaucratic skills also appeared pivotal in the configuration of the alternative agrifood system. Amy and Nick were a couple running a local a vegetable box-scheme. As they claimed, their experience in working for a local workers' cooperative was decisive in acquiring the knowledges and skills needed for setting their own agrifood business. These included not only did the 'know how' to package, buy and sell, but they also their 'know where' and 'know who'. As Nick said:

> I guess the most important part is buying. We were both buyers for [name of cooperative] so I would buy all the English veg, so I have a very good knowledge of who produces what in what part of the country and especially in the northwest of England ... So we knew what suppliers in what countries, where the good stuff comes from. So yes, so we had the knowledge, better knowledge, because [name of cooperative] are very specific about quality and I think through them we learned a lot about, not settling for anything less than very good. And we took that on for our own thing you know, and it's helped us to make it work smoothly. (Interview with representative of local box scheme 1)

In the case of this particular initiative, 'know who' has also been key for overcoming the absence of formal organic certification. Being a small-scale initiative, affordability of certification has been a challenge; however, good 'know who' from the side of the owners, in relation to both the suppliers and the users, has been pivotal for building as well as maintaining the relations of trust that would help them be 'organic' without certification. As Amy said:

> I think everyone who we deal with is certified. So we only buy from certified growers and producers. So we are slightly kind of hypocritical in that way but at least we know, we have the assurance, our customers have the assurance that all of our suppliers are certified. (Interview with representative of local box scheme 2)

In this context, communicative and social skills have also been pivotal, underlining the significance of the agrifood space as a 'social cognitive' space, based on as well as enhancing socio-spatial proximity between agrifood practitioners and the broader community. In many cases in Manchester, the agrifood practitioners' social and organisation skills also become central for the enhancement of the 'know who' and 'know how' of communities. In Manchester, this also becomes manifested in a number of educational courses, such as cookery classes, cook and taste activities, allotment-based mobile classrooms, all intended to 're-skill' communities by enhancing their situated knowledges and embodied socio-material experiences and engagement with and around food. In this way, following Giddens (1990), we could claim that agrifood practitioners' knowledges and skills become key for a gradual 're-skilling' of consumers, as will also be described below, which can help overcome the knowledge-expropriating mechanisms of the conventional agro-food chain.

Knowledge Production at the Consumption Site

Within the alternative agro-food system, consumption is configured as an integral part of understanding as well as configuring more sustainable agrifood supply chains. By consumption here, we refer to the broader processes and practices of accessing food that are possible for a diverse spectrum of communities. As already revealed from the above discussion, the alternative agro-food system also significantly relies on the diverse knowledges and skills developed by a broader spectrum of citizens and communities, usually also configured as the 'consumers' in the agrifood supply chain.

As seen above, in Cumbria, people's support has been essential for the future sustainability of the initiatives. The network of organic farmers under investigation have been attentive to the role of consumers as not only passive buyers of their products but also active citizens and agents of change, who could participate in public meetings, interact with local practitioners, and contribute to local decision-making. In this way, consumers' active engagement in agrifood networks and processes has also been pivotal for their configuration as knowing agents with a transformative potential, embedded not only in their enhanced knowledge and understanding of agrifood processes, but also of the broader spectrum of relations and values that exist and can be developed around food. In this context, a close connection between citizens' 'know how' and 'know who' is also important. The agrifood network space also constitutes a 'social cognitive' space where citizens 'know who' of local producers and retailers is pivotal for enhancing their 'know how' and 'know where', and therefore their support and further engagement in local agrifood processes.

The latter has been prevalent in a number of agro-food initiatives I visited in Cumbria. In the cases of a local farm shop and mill, the role of cafes has been key for enhancing consumers' 'personal experience through the senses' (Ingold 2000). During the interviews, I observed people of all ages watching the processes of milking, pasteurising, baking and so on. As Peter from the local farm and farm shop explained, enhancing consumers' knowledge about farming and food production processes has been essential for gaining their trust and support towards alternative agrifood practices. As also explained by Chris regarding their own farm and box scheme:

people come to the shop and see what we actually do and as a result purchase stuff. (Interview with local farm and box scheme)

Therefore, people's 'know who' and 'know how' appear to contribute to a broader re-skilling of consumers with regard to accessing food. Emily from a local organic farm raised the importance of direct contact with people. Once we finished our interview and after a short walk around the farm and her animals, we visited her shop. There were different pieces of meat and cheese, for which she was giving us elaborate information about its origin and taste, the different production and processing methods, while also offering tips and advice about how to cook it and relevant recipes, and giving us pieces to try it. We thus became witnesses of an alternative buying process, which was intrinsically embedded in the enhancement of consumers' situated knowledge of food, enhanced not only through the transfer of producers' personal knowledge about their food to the consumer, but also through the enabling of a more sensory experience of food, which, in many cases, as will be shown below also involved consumers more direct haptic engagement in not only tasting but also growing food.

The example of the social enterprise in rural Cumbria has been indicative towards such a direction. Following Toffler's definition of pro-sumers (1980), consumers' knowledge of agrifood practices and processes has been restricted not only to their more proximate socio-spatial relationship with producers and retailers, but also to their personal socio-material engagement and work with nature. Participating in educational and training courses on food growing has been pivotal for enhancing people's knowledge as well as appreciation of agrifood processes and practices, and therefore further fostering the creation of, what Kloppenburg et al. (1996) call 'insulated space' based on relations of solidarity and trust, reciprocity and mutuality between consumers, retailers and producers.

Processes of de-skilling and re-skilling of consumers have also been central in the alternative agrifood initiatives around Manchester. Re-skilling of consumers has been key in the vision of the local strategic partnership, employing food as a key means for supporting the well-being of local communities and economies. As Barbara from the partnership indicated:

a lot of people have no clue how to grow food. It's a mixture of getting people to rediscover lost skills through generations, but there are things about nutrition that we didn't know 50 years ago. It's not just about returning to time that we had a healthy diet ... We are trying to improve knowledge on what constitutes healthy diet, so that people can choose and have a balance in good health. (Interview with representative of local authority partnership)

Interestingly, in Barbara's understanding, re-skilling has been referring not only to the re-embodiment of those lost knowledges and skills, but also to those new ones that could contribute to the enactment of a healthier localised agrifood system. The connections between old and new knowledges and skills have also been evident in William's description of the interconnections between re-skilling and de-skilling. Involved in local community food growing projects, he explained that for him it is a process of re-skilling that is directly embedded in a process of gradual de-skilling, or as he calls it, a process of 'de-culturation' enacted through people's active engagement in food growing practices:

> people don't know how to cook, they don't know how to grow food, harvest it store it or anything ... so it's trying to affect the population, sort of de-culture them as well. (Interview with representative of urban agriculture network)

Taking such processes of knowledge production through deculturation further, as explained by the local authority strategic partnership representative, the latter is not only based on the re-acquisition of old and lost knowledges and skills, but also on the cultivation and adoption of new knowledges, which have not existed or have not been part of an old or traditional British food culture. As explained:

> It's easy to assume that we've lost some kind of food culture we had before, we've lost skills, but we have to recognize that it's something new that we try to develop as well. I don't think that Britain had a fantastic food culture that we lost and need to regain. There's a new [one]. (Interview with representative of local authority partnership)

Along these lines, several strategies have been employed as a way of shifting consumers' practices and knowledges. For example, in one of Manchester's workers' co-operatives, people get involved in a diversity of social relations and interactions, which go beyond the narrow impersonal economic transactions of the conventional agro-food retailing system. Like the Athenian agora, the market is simultaneously a space for diverse forms of sociality—for personal relations, for the reproduction of community, for the exchange of knowledge and opinion, but also for political action (Horton 2003). Consumption practices become a manifestation of the wider ethics of solidarity and reciprocity to the producers and retailers involved in the envisioning of an alternative agro-food system. For communities involved, the practice of consumption becomes a political act, based on the exchange of not only goods and services, but also on sharing of knowledges and ideas around the agro-food production system. A socio-spatial proximity and development of personal relations with the workers of the co-op become means for a wider transfer of knowledge and sharing around the origin of the food, its quality, its properties and the social values embedded in it. Producers and consumers engage in an array of practical operations and tasks which draw on, extend and share their knowledge about the practicalities, ethics and politics of agro-food production, distribution and consumption (Psarikidou and Szerszynski 2012). Therefore consumption becomes part of a wider experience for the exchange of knowledge and personal experiences about fruits and vegetables, their processes of preparation and cooking.

However, at the same time, alternative agrifood spaces are also transformed into important social cognitive spaces for local communities. Both Manchester's allotments and community gardens have been illustrative of the ways urban growing spaces have been important spaces for growing local communities' social skills and relations with each other. Michael, the representative of Manchester's allotments society, approached allotments as 'community spaces', encouraging us to consider the diverse types of sociality that can be developed around people's socio-material engagement with food. My visit at a local urban agriculture project has also been illustrative of the diverse processes of re-skilling, knowing and learning attributed to the social relations developed around growing food. As described earlier, digging, moving old trees, planting new trees and other species were all different growing practices in which people of different ages, ethnic backgrounds and level of experience were involved. Such practices have therefore been important for configuring agrifood spaces as spaces for collective knowing and learning, facilitated through the social co-operative relations amongst local people. People have been working in groups, exchanging knowledges and ideas about growing as well as cooking, but also sharing broader concerns and questions around food security, access and safety. At one side of the forest garden there was a tent with several vegan dishes cooked by the forest garden community. At this particular place, people's food sharing experience has been pivotal for enhancing not only their 'know who' but also their 'know how' and realising the significance of the connections between the two. However, in all these different ways, the garden has also been important for configuring as well as prefiguring new social spaces of collective action through learning, which would lead to a broader food system transformation based on relations of solidarity, justice and inclusion.

The latter has also been prevalent in the work of a local social enterprise working with the so-called 'disadvantaged' communities. As discussed in their report, for them, food has been central for configuring more socially just and inclusive societies, particularly through the cultivation of social skills, both of those vulnerable parts of the population as well as those who are supporting their agrifood practices. In this context, diverse communities' engagement with and around food processes, both as growers, distributors as well as consumers, becomes important for the reconnection between those different parts of the local population through and around food.

Based on all above, we could claim that, in both cases of Cumbria and Manchester, community agrifood spaces become important spaces of knowing for food system transformation. They constitute important 'social cognitive' spaces, where there is a direct and mutual co-production between communities' 'know how' and 'know who', helping local communities to not only reconnect with agrifood practitioners and processes, but also develop a feeling of common belonging, relations of solidarity, reciprocity and trust with proximal and distant others. Here, as already seen, processes of de-skilling and re-skilling—with regard to buying, growing and cooking—are very important for reconfiguring everyday knowledges and practices around food. In this context, such alternative agrifood spaces are transformed not only into cognitive, but also a 'social cognitive' space in which local communities' 'know how' contributes to the acquisition of not only a better knowledge about and re-connection with alternative agrifood processes, but also with each other. In this way, the so-called 'consumers' are not only passive recipients, but, through their consumption practices, they become significant allies for the configuration of an alternative agro-food supply system, as well as active agents for the enactment of more proximate social relations and skills, within as well as beyond the agrifood system.

Alternative Agro-Food Networks and the Knowledge Economy

Knowledge Production in AAFNs

The above analysis underlined the centrality of knowledge in configuring alternative agro-food systems in ways that challenge existing knowledge hierarchies within conventional agrifood systems. Within AAFNs, different forms of knowledge appear to variously interchange and co-exist, transcending beyond any traditional dichotomies between scientific and other knowledges, based on alternative conceptualisations and understandings of scientific as the dominant form of knowledge and its relation with other knowledges.

As the above description reveals, science still plays a key role in the configuration of the alternative agro-food knowledge system. However, not only does it provide the foundations for configuring an 'alternative to science' agro-food model, but, it also becomes an integral part in the development of an alternative 'science-based' agrifood model. Following Scott's conceptualisation of 'technē'[5] (1998), scientific knowledge for alternatives does not aim to invent or construct new patterns of knowledge, neither does it aim to standardise knowledge. In many cases, like with the organics, it is an explicit form of knowledge that is based on tacit, experience-based knowledge (what Scott calls 'mētis'), and, in doing so, it aims to both explain natural phenomena and facilitate the communication of knowledge about them (Scott 1998). As Polanyi would argue, it brings us back to the understanding of scientific knowledge which is always rooted in tacit knowledge, constituting a means for the externalisation (Nonaka 1994) as well as better communication of tacit knowledge.

It can thus be argued that, in AAFNs, scientific knowledge constitutes part of a unified knowledge system (see Turnbull and Verran 1995) that goes beyond the mainstream understanding of science as a highly rationalised and technocratic process. Scientific knowledge is thus configured as a particular type and stage in a broader process of knowledge production, which is rooted in agrifood practitioners' situated knowledges based on

---

[5] According to Scott (1998), despite the fact that 'techne' could resemble the self-image of modern science, it is actually codified, verified and documented knowledge, a set of systematic and impersonal rules, which aim to comprehensively document and transfer the already existing knowledge (p. 320).

their broader socio-material engagement in agrifood processes. It constitutes a form of knowledge which is based on the observation of spaces, but can also bring into existence new cognitive spaces which can expand specific geographically bounded localities to configure new agrifood knowledge systems. It is the explicit form of knowledge which helps create new social spaces of communication, interaction and exchange of knowledge between different kinds of experts—scientific and local experts—and where the 'know what' (of the scientists) facilitates the transfer of the 'know how' (of the practitioners) through the cultivation of the 'know who' (of both scientists and practitioners). In this context, it offers an alternative conceptualisation of scientific knowledge which is based on its embedding in complex social and natural environments and a greater diversity of experts co-exist for the production of agrifood knowledge (see also Callon et al. 2009).

In this context, within AAFNs, agrifood knowledge production is configured as an ongoing learning process acquired through direct involvement with land, food and nature (Ingold 2000). For most of the practitioners, agrifood knowledge has still been situated in people's experiences, interactions and entanglements in particular socio-material environments, whereas any explicit or codified forms of knowledge derive from such complex socio-material engagements within natural and human environments (Scott 1998; Callon et al. 2009; Latour 1986, 1987; Turnbull and Verran 1995). Therefore, for most of the practitioners, agrifood production processes are mainly configured as ongoing learning processes resembling Scott's conceptualisation of 'mētis' (1998)—that is, where 'know how', the practical wisdom of long experience as well as personal observation and experimentation with nature—or else what is also described as the local or situated knowledge—is equally important for the production of new knowledge for agriculture and food. New science models are conceptualised which go beyond the mainstream understanding of science to embed it in a more dialectic relationship with other knowledges as well as between different types of experts and their engagements within wider social and natural environments (see e.g. Polanyi 1967; Callon et al. 2009; Turnbull and Verran 1995).

As seen in the previous sections, tacit, local, embodied and embedded knowledges all variously engage and interact for the configuration of an alternative agrifood knowledge system. Such plural processes of knowing and learning do not only prevail at the production, but also the agrifood distribution and consumption sites, underlining the diversity of experts as knowing agents for agrifood change. In these cases, knowledges take the form of mutable knowledges, constantly adapting to changeable socio-natural environments. In this context, the agrifood landscapes take the form of taskscapes (Ingold 2000), where diverse practical operations and tasks—usually conducted by skilled agents—'get their meaning from their position in an ensemble of tasks, performed in series or in parallel, and usually by many people working together' (Ingold 2000, p. 195). And, the agrifood practitioners' knowledge is attached not only to their interaction with the land, but also to the more diverse spectrum of social interactions and exchanges of knowledge that take place around the land. It thus becomes the social space that is built on the complexity of interactions and the diversity of knowledges that become both embedded and embodied as part of the agrifood practitioners' engagement and work with nature (Ingold 2000).

In this context, knowledge production for the alternative agro-food sector could be claimed to contribute to a 'resurrection' of old knowledges and skills, based on processes of re-skilling and de-skilling. On the one hand, a gradual de-skilling becomes important for agrifood communities' alienation from the dominant agrifood knowledges and skills of the conventional agrifood system. On the other hand, a re-skilling process also becomes important for reconnecting alternative agrifood communities with those lost or forgotten, usually described as old or forgotten, knowledges and skills that become pivotal for the configuration of a new agrifood knowledge model.

However, as discussed above, in many cases, knowledge production within alternative agrifood systems is also the outcome of a combination of both old and new knowledges, the latter coming from the recombination and adaptation of existing forms of knowledge. In this context, new forms of explicit and tacit, codified, embedded and embodied knowledges are developed, combined, as well as converted through diverse externalisation and internalisation processes. New experts appear within such complex agrifood knowing spaces, including scientists, producers, retailers and consumers. In this way, existing knowledge hierarchies are overcome, based on a plurality of knowledges and experts that become important for the production of new knowledge for the agrifood system. For example, in the case of the producers, new IT-based communication and marketing skills are configured as important for empowering alternative agrifood stakeholders everyday practices. Consumers appear as a new diverse community of knowing agents, the new experts (Collins and Evans 2002), variously engaging in agrifood spaces and practices, and actively contributing to the production as well as sharing of knowledge for configuring alternative agrifood systems. In this context, the alternative agro-food landscape constitutes not only a cognitive but also a social cognitive space, where several knowledge conversion processes—the internalisation and externalisation of different knowledges and skills—are grounded on the development of stronger co-operative links and networks, which succeed in facilitating the 'de-skilling' and 're-skilling' of engaged actors, producers, retailers and consumers. It is the space, which by enabling a greater socio-spatial proximity between a diversity of experts, becomes pivotal for the exchange, sharing and transfer of plural knowledges and skills that are central for the configuration of the alternative agrifood system as an agrifood knowledge system.

Knowledge Production in AAFNs and the Knowledge Economy: A Comparative Perspective

Based on this description, we could identify several elements through which the knowledge production of the alternative agro-food sector could resemble the one of the knowledge economy.

As we have already seen in previous chapters, the spark of innovation in the knowledge economy relies on a continuous dialogue and interplays between different types of knowledge; despite the obvious prioritisation of techno-scientific knowledge, explicit and tacit—embrained and encoded, embedded and embodied—forms of knowledge seem to complement each other in the production of new knowledge and innovation. Knowledge production is conceived of as a continuous learning process, which does not so much rely on the discovery of new knowledge, but on the recombination and adaptation of existing forms of knowledge. Therefore, the knowledge economy becomes the outcome of a series of knowledge conversion processes— for example internalisation and externalisation processes—which also underline the significance of social interactions and relations mobilised around it. The knowledge production process is thus constructed around a 'social space of knowing' (Amin and Cohedent 2004), which provides a fertile ground for the generation of knowledge through communication. Innovation seems to build upon 'communities of practice' (Knorr Cetina 1981), where new knowledge appears to stem from routine activities, social interactions and engagements in an array of conversations, encounters, scripts, memories, routines and stories (Callon

1999; Latour 1986). And, in this framework, the practitioners' tacit knowledges and practical experiences, as well as knowledge sharing among communities of practice acquire a central role in the production of a new knowledge driven by innovation (for further analysis, see also e.g. Smith 2000; Nonaka 1994; Blackler 1995; Birch 2007; Lam 2000; Amin and Cohedent 2004).

In the above synopsis, we could identify several indications which could lead us to approach alternative agro-food economies as knowledge economies. As observed in this chapter, knowledge production in alternative agro-food economies constitutes an ongoing learning process, where explicit and tacit, expert and lay, old and new, encoded, embedded and embodied forms of knowledge co-exist for the production of a 'radical innovation' (Dosi 1988; Johnson 1992), which could stem not only from the 'forgetting' and 'creative destruction' of old knowledges, and the resurrection of local, context-dependent knowledges (cf. Dosi 1988; Dosi et al. 2008), but also from the recombination and adaptation of a more diverse spectrum of existing knowledges—including scientific knowledge and ICT. As we have already seen, knowledge production in the agro-food sector focuses not on the rejection of scientific knowledge, but on a fruitful re-conceptualisation and utilisation of scientific knowledge for the creation of an alternative knowledge-based agro-food system. It does not solely rely on the resurrection of old knowledges and skills, but also involves the adaptation of emerging techno-scientific knowledges and skills, linked to information technologies (ITs). However, it goes beyond a prioritisation of these knowledges, and transforms them into parts of a more diverse knowledge production process based on the co-existence and interplay between a more diverse set of knowledges. The new knowledge is not an outcome of a truly scientific discovery but a 'work in progress' (Ingold 2000; Franklin 2002), dependent on the tacit and situated knowledges attached to diverse knowing agents' personal engagement, bodily experiences, interactions and experimentation with the human and non-human others (Ingold 2000; Haraway 1991). It is a knowing process that, like within other knowledge economies, tacit ways of knowing are important for the production of new knowledges and a knowledge system that is based on diverse knowledge conversion processes through which agrifood knowledge gets communicated, transferred and shared across a plurality of equally important experts involved in the production, distribution and consumption of food. In this way, the agro-food knowledge space is also transformed into a social cognitive space, where people's social and communicative skills appear pivotal for enabling processes of knowledge conversion through knowledge sharing, but also for pluralising the types of knowledges and experts that can contribute to the production of new knowledge for the agrifood system.

Based on the above description, we could identify several similarities which could lead us to approach the alternative agro-food sector as a knowledge economic sector. The co-existence of scientific and tacit forms of knowledge, the knowledge interplay, transfer and sharing, the knowledge conversion processes, as well as the transformation of the knowledge space into a social space of knowledge, can all lead us to argue for a possible re-thinking of the alternative agrifood economy as a knowledge economy, or even a knowledge bioeconomy, whose knowledge production significantly lies on the production of new knowledge that builds on a constant socio-material engagement with life and nature.

However, there is also a range of differences between these sectors, which are mainly embedded in the highly capitalocentric economic discourse built around the KBE, KBBE and agriculture. As already discussed in previous chapters, the knowledge economic and bio-economic narratives are grounded on the promissory premises of profit maximisation and ever-increasing productivity. In their pursuit of global competitive economic advantage, they depend upon the transformation of knowledge and biological life into marketable units, commodities which can be produced, distributed, sold and monopolised. In this way, they renovate capitalism by moving beyond the natural limits to capital introduced by earlier economic forms. Accumulation of surplus value increasingly relies on the productive forces of 'immaterial labour' and the production and commodification of immaterial goods and services. In this context, knowledge becomes central, not only as a factor of production leading to the dematerialisation of goods and services to be sold for profit, but also as a commodity that can be appropriated, produced, packaged and sold as a piece of information for profit. And, here, biological life appears key: not only as a resource but also as a carrier of knowledge and a 'piece of information' (Thacker 2005), which can get decontextualised, reified, made and commodified in ways that can contribute to the enhancement of its own productivity (Cooper 2008) for the accumulation of capital and surplus value.

The above description provides significant clues that can help us approach the AAFNs KB(B)E as an alternative to the mainstream KBBE but also an alternative KBBE. In many ways, the alternative agrifood knowledge practices appear to move beyond the mainstream capitalocentric discourse (Gibson-Graham 1996, 2006) of profit maximisation and ever-increasing productivity portrayed within the dominant agrifood KBBE. Knowledge and nature are not seen as resources for the reproduction of capitalism. The production of new knowledge is not related to a future potential of transforming nature. It stems from a continuous experimentation, personal observation and work with nature, its biological patterns and limits to capital. Knowledge and life remain recommodified, part of a process of sharing, whose surplus value transcends narrow understandings of 'the economic' and 'economic value'. Configured as social cognitive spaces, alternative agrifood spaces re-embed knowledge production in the social relations of solidarity, co-operation and connectivity developed around nature. In doing so, it re-embeds knowledge production in the broader spectrum of social and cultural values and symbolic meanings attributed to agriculture and food, land and nature (see Graeber 2001). Within this context, the alternative agro-food sector's 'immaterial labour' (see Lazzarato 1996; Hardt 1999) moves beyond the manipulation of knowledge and information, as well as the creation of human affect, values and relationships with the purpose of capital accumulation. The alternative agro-food sector's 'immaterial labour' embeds labour and knowledge production in a range of decommodified social relations, symbolic meanings and values attributed to agriculture and food, opening up a wider understanding of 'the economic' (Gibson-Graham 1996, 2006) and encouraging us to not only re-think but also reclaim the concept of the 'knowledge economy'.

References

Amin, A. and Cohedent, P. (2004) *Architectures of Knowledge: Firms, Capabilities and Communities.* Oxford: Oxford University Press.

Birch, K. (2007) Knowledge, Place and Power: Conceptualising Value Creation in Knowledge-Based Commodity Chains. Working paper. Centre for Public Policy for Regions. University of Glasgow.

Blackler, F. (1995) Knowledge, Knowledge Work and Organizations: An Overview and Interpretation. *Organisation Studies*, 16(6): 1021–1046.

Brown, J.S. and Duguid, P. (1991) Organizational learning and communities of practice: towards a unified view of working, learning and innovation, *Organisation Science* 2(1):40–57.

Callon, M. (1999) The Role of Lay People in the Production and Dissemination of Scientific Knowledge. *Science, Technology and Society* 4(1):81–94.

Callon, M., Lascoumes, P. and Barthe, Y. (2009) *Acting in an uncertain world: An essay on technical democracy*. Translated from French by Graham Burchell. Cambridge, MA, USA, London: MIT Press.

Collins, H.M. and Evans, R. (2002) The Third Wave of Science Studies: Studies of Expertise and Experience, *Social Studies of Science* 32(2):235-296.

Cooper, M. (2008) *Life as Surplus: Biotechnology and Capitalism in the Neoliberal Era*. Seattle: University of Washington Press.

Davis, J. and Hinshaw, K. (1957) *Farmer in a Business Suit*. New York: Simon and Schuster, Inc.

Dosi, G., 1988. The nature of the innovative process. In: Dosi, G. et al. (Eds.) Technical Change and Economic Theory. London: Pinter: pp. 221-238.

Dosi, G., Faille, M. and Marengo, L. (2008) Organisational Capabilities, patterns of knowledge accumulation and Governance Structures in Business Firms: An Introduction. *Organisational Studies* 29: 1165-1185.

Epstein, S. (2007) *Inclusion: The Politics of Difference in Medical Research*. Chicago: University of Chicago Press.

Fonte, M. and Grando, S. (2006) A Local Habitation and a Name: Local Food and Knowledge Dynamics in Sustainable Rural Development, *CORASON project*. Available online at www.corason.hu.

Franklin, A. (2002) *Nature and Social Theory*. London: Sage.

Gibson-Graham, J.K. (1996) *The End of Capitalism (As We Knew It): A feminist Critique of Political Economy*. Oxford UK and Cambridge USA: Blackwell Publishers.

Gibson-Graham, J. K. (2006) *A Postcapitalist Politics*. Minneapolis, MN: University of Minnesota Press.

Giddens, A. (1990) *The Consequences of Modernity*. Cambridge: Polity Press.

Giddens, A. (1991) *Modernity and Self-Identity: Self and Society in the Late Modern Age*. CA: Stanford University Press.

Graeber, D. (2001) *Toward an Anthropological Theory of Value*. Palgrave Macmillan.

Haraway, D. (1991) 'Situated knowledges: The science question in Feminism and the privilege of partial perspective', in Haraway, D. (ed.) *Simians, Cyborgs and Women: The Reinvention of Nature*. New York: Routledge.

Hardt, M. (1999) Affective Labour, *Boundary 2* 26(2): 89-100.

Horton, D. (2003) Green distinctions: the performance of identity among environmental activists. In Szerszynski, B., Heim, W. and Waterton, C. (eds.) *Nature performed: Environment, Culture and performance*. Cambridge: Blackwell: pp. 63-77.

Ingold, T. (2000) *The perception of the environment: essays on livelihood, dwelling and skill*. London: Routledge.

Johnson, B. (1992) Institutional learning. In Lundvall, B. (Ed.), *National Systems of Innovation*. London: Pinter.

Kloppenburg, J., Henrickson, J. and Stevenson, G.W. (1996) Coming in to the foodshed, *Agriculture and Human Values* 13:33–42.
Knorr Cetina, K. (1981) *The Manufacture of Knowledge*. Oxford: Pergamon Press.
Lam, A. (2000) Tacit Knowledge, Organisational learning and Societal Institutions: An Integrated Framework, *Organization Studies* 21(3):487–513.
Latour, B. (1986) Visualisation and Cognition: Thinking with Eyes and Hands, *Knowledge and Society* 6:1–40.
Latour, B. (1987) *Science in action*. Cambridge, MA: Harvard University Press.
Latour, B. (1993) *We Have Never Been Modern*. Cambridge, MA: Harvard University Press.
Lazzarato, M. (1996) Immaterial Labor. In Hardt, M. and Virno, P. (eds.) *Radical Thought in Italy: A Potential Politics*, Minneapolis: University of Minnesota: pp. 133–50.
Nonaka, I. (1994) A Dynamic Theory of Organisational Knowledge Creation, *Organisation Science* 5(1):14–37.
Permaculture Association (2009) Knowledge base. Available online at http://www.permaculture.org.uk/.
Permaculture Association (2011a) The basics. Available online at http://www.permaculture.org.uk/.
Permaculture Association (2011b) *What is Permaculture*. Available online at http://www.permaculture.org.uk/.
Polanyi, M. (1967) *The Tacit Dimension*. New York: Anchor Books.
Psarikidou, K. and Szerszynski, B. (2012) Growing the Social: Alternative agro-food networks and social sustainability in the urban ethical foodscape, *Sustainability: Science, Practice and Policy* 8(1):30–39.
Royal Horticultural Society (2011) Gardening. Fleece. Available online at http://www.rhs.org.uk/.
Scott, J.C. (1998) *Seeing Like a State. How Certain Schemes to Improve Human Condition Have Failed*. Yale University Press.
Smith, K. (2000) What is the 'knowledge economy'? Knowledge-intensive industries and distributed knowledge bases. Paper presented to DRUID Summer Conference on The Learning Economy—Firms, Regions and Nation Specific Institutions. June 15–17, 2000.
Soil Association (2011a) What is organic. Available online at http://www.soilassociation.org/.
Soil Association (2011b) Organic Standards. Available online at http://www.soilassociation.org/whatisorganic/organicstandards.
Soil Association (2011c) Soil Association Organic Standards: Farming and Growing. Available online at http://www.soilassociation.org/LinkClick.aspx?fileticket=l-LqUg6iIlo%3d&tabid=353.
Thacker, E. (2005) *The Global Genome: Biotechnology, Politics and Culture*. Cambridge, MA: MIT Press.

Toffler, A. (1980) *The third wave*. Bantam Books.
Turnbull, D. (1997) Reframing knowledge and other local knowledge traditions. *Futures*, 29(6): 551–562.
Turnbull, D. and Verran, S. (1995) Science and Other Indigenous Knowledges in Sheila Jasanoff et al. (eds.), *Handbook of Science and Technology Studies*, London/Thousand Oaks, Sage (revised edition 2002): pp. 115–139.
Wilsdon, J., Wynne, B. and Stilgoe, J. (2005) The Public Value of Science. Or how to ensure that Science really matters. *Demos* Available online at http://www.demos.co.uk/publications/publicvalueofscience. Retrieved May 25, 2009.

CHAPTER 6

Conclusions: Reclaiming the Knowledge Economy

THE ARGUMENT SO FAR

Throughout this book, I have aimed to bring into creative dialogue two contemporary and allegedly contradictory developments: the rise of alternative agro-food networks (AAFNs) and the vision of a knowledge-based (bio)-economy (KB(B)E). In doing so, I attempted to unfold the different dimensions in which an AAFN economy could relate to the KB(B)E. I explored the nature of these two developments and the narratives around them, and tried to unpack the particular knowledges and knowledge production processes which could help us develop a better understanding of the ways an AAFN economy could not only be approached as 'an alternative to' the KB(B)E, but also constitute an 'alternative' KB(B)E.

The first chapter provided some first grounds justifying an exploration of the potential links between AAFNs and the KB(B)E. By offering a historical and political economic approach to the KB(B)E, it offered insights into the role of knowledge, nature and labour processes in the formulation of a new face and phase of capitalism (Sunder Rajan 2006). It underlined the dematerialisation processes involved in modern relations of production, the commodification of knowledge and information, the manipulation of symbols, human affects and emotions, the switch from material to 'immaterial' and 'affective' forms of labour (Lazzarato 1996; Hardt 1999; Morini 2007), and underlined their significance for accumulating capital through the transformation of nature into not only a commodity but also into variable capital, a productive labour force which can go beyond the human and natural limits to 'productivity' (Brennan 2000; Thacker 2005).

In doing so, it highlighted those particular elements and ways that the KB(B)E has currently come to constitute a master economic narrative which is subordinate to a specific political economic vision, as well as research and policy innovation agendas, which, in the promise of certain ideas of 'productivity' and 'growth' (Sunder Rajan 2006; Cooper 2008; Wallace 2010), result in re-producing and renovating capitalism, while marginalising other research and policy options for sustainability.

Following up from this rationale, in the second chapter of the book, I aimed to unpack the different ways in which an agro-food economy can be seen as a knowledge economy. I explored the different knowledges and knowledge production processes involved in both, and identified the similarities and differences, as well as the final common threads, which could encourage us to approach an agro-food economy as a knowledge economy. As revealed, within the KB(B)E, a co-existence between different forms of knowledge, both explicit and tacit—for example, embrained and embodied, encoded and embedded (see Nonaka and Takeuchi 1995; Collins 1993; Blackler 1995; Lam 2000)—seems pivotal for understanding the complex and diverse knowledge dynamics that take place for the production of innovation. The recombination and adaptation of these knowledges through multiple conversion processes—for example, internalisation, externalisation, familiarisation, socialisation (see Nonaka 1994)—are important for understanding such 'knowing spaces' (Amin and Cohedent 2004) as 'social cognitive spaces', where the production of new knowledge is part of a 'learning by doing' process (Lam 2000) majorly developed within 'communities of practice' (Knorr Cetina 1981). In this context, innovation is configured as a product of a continuous communication, interaction and interplay between different types of communities, or else, as Callon (1999) and Latour (1986) would have argued, a product of connections between previously unconnected and heterogeneous ingredients, an outcome of learning through doing and socio-material engagements acquired within routines, conversations, meetings, scripts memories and stories.

The latter is also prevalent within agro-food knowledge production processes. As revealed, tacit and explicit forms of knowledge appear to variously interchange and co-exist, configuring agro-food spaces as 'knowing spaces' based on constant interactions between material and immaterial labour and the creative dialogue among diverse communities of knowing agents and experts. Following Hardt and Negri (2004), despite the apparent material aspects of their labour, farmers are 'knowledge workers' whose labour is based on as well as contributing to the production of knowledge and innovation typical of other forms of 'immaterial labour', involving skill, affect and judgement. At the same time, agricultural processes are also increasingly based on the emergence and engagement of new 'knowledge workers' extending agro-food activities and knowledge practices beyond the farm level. Such observations encourage us to think of the diverse processes of dematerialisation taking place in the agro-food labouring processes that can help us realise the ways agro-food economies can be part of a mainstream, growth-oriented capitalocentric vision of the knowledge economy, but also consider the diverse ways or possibilities of diverting from it. It encouraged us to realise the centrality of knowledge within an agro-food economy, and therefore further investigate the possibility of an alternative agro-food knowledge economy.

Based on my findings from two networks of diverse types of alternative agro-food initiatives in Northwest England—the rural county of Cumbria and the city of Manchester—I looked into the different knowledges, skills, and knowledge production processes involved in their agrifood practices, and explored and their role in configuring an alternative agro-food knowledge economy. As revealed from the analysis, the AAFN economy incorporates several characteristics which could lead us approach it as a KB(B)E. Its 'radical innovation' (Dosi 1988; Johnson 1992) relies not only on the 'creative destruction' of old knowledges and the resurrection of local context-dependent knowledges (cf. Dosi et al. 1993), but also on a fruitful dialogue, recombination and adaptation of different forms of knowledge—including scientific knowledge, modern technical skills, managerial, marketing skills and IT. Scientific and tacit, old and new, local and situated knowledges seem to co-exist and interact in configuring an alternative knowledge-based agro-food economic model. Through a diversity of knowledge conversion processes, agrifood knowledge appears to be the outcome of a continuous knowing process developed around diverse agrifood knowing agents' social and communicative skills. In this context, within AAFNs, agrifood innovation does not stem from an absolute rejection but a constructive reconfiguration of science based on its fruitful recombination with a more plural spectrum of equally important knowledges of a diverse community of experts (Amin and Cohedent 2004). As with KB(B)E knowledge production, the alternative agro-food knowing space constitutes a 'social cognitive space', which, is not just based on rejection of techno-scientific knowledge, but a fruitful re-conceptualisation and engagement with it as part of more plural process of knowledge production based on a fruitful dialogue between a more diverse spectrum of knowledges and skills acquired through the agro-food knowledge workers' socio-material engagement and experimentation, embodied experiences and interactions with human and non-human others (Ingold 2000; Haraway 1991).

However, the AAFN knowledge economy also appeared to carry some particular characteristics which differentiated it from the mainstream capitalocentric discourse of the KB(B)E. As discussed already, the KB(B)E is primarily configured around a process of accumulation of capital and surplus value through the commodification and appropriation of the productive potential of knowledge and life. In contrast, the alternative agro-food knowledge economy seemed to go beyond the dominant capitalocentric discourse of profit maximisation and ever-increasing productivity. Working with respect to the biological patterns of nature and limits to capital, within the AAFN KB(B)E, agrifood knowledge has been configured as a communal resource, maintained, produced and acquired around social relations of collective work, solidarity and sharing between a diversity of experts and their 'communities of practice' (Knorr Cetina 1981). Going beyond narrow ideas of 'productivity' (cf. Brennan 2000; Thacker 2005), it appeared being based on de-commodified experiences, relations and understandings of nature. Through their engagement in a wide array of social relations of knowledge sharing, the AAFNs' immaterial and affective labour contributed to the attribution of a broader set of socio-cultural meanings and values to agro-food knowledge practices and relations of production. Based on relations of co-operation, solidarity and trust, the AAFN KB(B)E was configured as the space for the reproduction of the community and an expression of care towards proximal and distant human and non-human others, and in which food, knowledge and ideas become objects of decommodified relations of exchange and a 'sign value' (Baudrillard 1972) related to wider social goals and relationships, socio-material entanglements and work with nature. In this context, the AAFN KBBE also gets a wider political meaning, in which the 'knowing processes' contribute to the creation of a space for a 'purposive political acts' and a 'life politics' (Giddens 1991) performed by a diversity of knowing agents, who, through their engagement in alternative agrifood practices, aim at a transformation within the agrifood system and beyond, moving beyond narrow capitalocentric approaches to 'the economic' and hegemonic understandings of the 'knowledge economy'.

Reclaiming the Knowledge Economy: The Agro-Food Sector and Beyond

The knowledge economy has currently come to constitute a master economic narrative which is subordinate to a specific political economic vision, as well as research and policy innovation agendas that, in the promise of certain ideas of 'productivity' and 'growth' (Sunder Rajan 2006; Cooper 2008; Wallace 2010), result in re-producing and renovating capitalism while marginalising other research and policy options for sustainability.

Our investigation of the AAFN economy as a knowledge economy encourages us to consider those other knowledge economies that currently exist but remain marginalised within dominant research, policy and innovation agendas for agri-food sustainability. It points our attention to the potential of the AAFN economies to configure as well as enact alternative knowledge economies beyond narrow capitalocentric understandings of 'the economic' and dominant techno-scientific approaches to 'agrifood innovation'. It thus encourages us to realise the innovation potential that is embedded in all agrifood knowledge systems, and appreciate the plurality of knowledges and agrifood experts that are important for the production of new knowledge and innovation for agrifood sustainability. It helps understand the AAFN economy as an alternative, more plural and inclusive, knowledge economy based on the knowledges and expertise of a broader spectrum of actors from farm to fork. But it also encourages to configure the knowledge economy in more plural and inclusive ways: not only by considering the plurality of knowledges, processes and agents of knowing that are important for the production of innovation within the knowledge economy, but also by realising the diversity of knowledge economies that need to be taken into consideration when configuring as well as enacting future agrifood research, policy and innovation trajectories for agrifood sustainability.

Following Gibson-Graham's (1996, 2006) post-capitalist iceberg economy, our investigation of the AAFN knowledge economy encourages us to think of the 'knowledge economy' as a diverse economic landscape of cohabitation and contestation between different 'knowledge economic' forms that are all equally important and interdependent in configuring more sustainable agro-food futures. It encourages us to realise the possibility of pluralising 'knowledge economies': to move beyond a singular and narrow capitalocentric and technocentric approach to the knowledge economy, and consider these other 'knowledge economies' that currently exist but remain silenced from today's research and policy agendas for agrifood sustainability. It therefore contributes to widening the identity of the 'knowledge economic' to include all of those practices that are currently excluded or marginalised due to the naturalised presumption of a capitalist techno-scientific hegemony. Following Gibson-Graham's work (1996, 2006), the AAFN configuration of a knowledge economy contributes to signalling the possibility of re-thinking as well as re-making the knowledge economy in alternative ways; for realising the possibility of a more plural and inclusive knowledge economy, encompassing a more diverse spectrum of knowledges and knowing agents for agrifood innovation; for making space for new knowledge economic becomings that can contribute to more democratic configurations as well as practices and policies of innovation beyond established agrifood knowledge hierarchies and power relations supporting narrow techno-solutionist approaches to innovation.

Such understandings are also important for reclaiming the knowledge economy within as well as beyond the agro-food sector. They encourage us to consider all those other, currently defined, 'knowledge economic' sectors—from Biomedicine to Information and Communication Technologies (ICTs), from Higher Education Institutions (HEIs) to Health and the Creative Industries—and reconfigure them as diverse knowledge economic landscapes based on plural knowledges and experts, as well as ideas of 'innovation', 'productivity' and 'growth'. They therefore provide as well as call for a framework for understanding as well as enacting the 'knowledge economy' in more plural and inclusive ways: by taking into consideration the knowledges, voices and standpoints of a broader spectrum of experts; by extending the idea of 'experts' to include scientists as well as those other knowing agents whose diverse standpoints and knowledges, tacit and explicit, old and new, traditional and modern, situated and local, become equally important for informing as well as configuring science and innovation in more inclusive ways; by configuring the knowledge economy as a hybrid knowledge system, based on more complex interactions and co-production processes between those different knowledges, but also more plural relationships to science and technology; by considering 'the knowledge economy' in plural, and thus realising and opening up the innovation potential of all those usually configured as 'alternative' practices and agents of knowing—from alternative medicine to makerspaces, from urban food growing to bike-sharing and community renewable energy schemes—that usually remain marginalised due to dominant research, policy and innovation trajectories and master narratives for the 'knowledge economy'.

Therefore, a critical re-thinking of the knowledge economy can contribute to the empowerment of those diverse knowledge economic developments that are currently affected by narrow capitalocentric and technocentric understandings of the term. Such a move will be pivotal for not only acknowledging the innovation potential of such economic organisations, but also for re-configuring research and policy in ways that would support the enactment of such alternative knowledge economic realities for innovation and growth. Reclaiming the knowledge economy is key for pluralising research and policy agendas in ways that would help not only the enactment of those currently configured as 'alternative' knowledge economies, but also the realisation of the possibility of a more diverse economic landscape for innovation and growth.

However, such understandings can also be important for challenging ideas of 'alternative' within the alternative agrifood as well as other knowledge economies. In order to build a new kind of knowledge economic reality, we need to overcome current hierarchical orderings within the knowledge economy. Reconfiguring the knowledge economy as a diverse economic landscape entails dismissing ideas of 'the alternative' as subordinate or oppositional (see also Allen et al. 2003). It entails the forgetting of pre-established stereotypes or assumptions of what or who is 'the powerful' or 'the superior' (Gibson-Graham 1996). And, for this to be possible, not only do we need to develop a richer language of 'the knowledge economy', but also we need to reconfigure our own subjectivities as active agents in new collectivities that can desire and enact economic realities that can displace binaries and singular visions of 'innovation' and 'growth' within the knowledge economy.

Building a new knowledge economic reality is possible. However, it entails a collective re-thinking and re-doing of the 'knowledge economy'. Alternative Agro-Food Networks provide an interesting space for achieving that: for not only realising the possibility and need for conceptualising and enacting the 'knowledge economy' in more inclusive ways, but also organising the agrifood collectivities that are needed (and we should all be part of) in order to realise more plural research, policy and innovation pathways for agrifood sustainability, properly conceived.

References

Allen P., FitzSimmons M., Goodman M. and Warner K. (2003) Shifting plates in the agrifood landscape: the tectonics of alternative agrifood initiatives in California, *Journal of Rural Studies* 19(1): 61–75.

Amin, A. and Cohedent, P. (2004) *Architectures of Knowledge: Firms, Capabilities and Communities.* Oxford: Oxford University Press.

Baudrillard, J. (1972) *For a critique of the political economy of the sign.* St. Louis: Telos.

Blackler, F. (1995) Knowledge, Knowledge Work and Organizations: An Overview and Interpretation. *Organisation Studies*, 16(6): 1021–1046.

Brennan, T. (2000) *Exhausting Modernity: Grounds for a new economy.* London and New York: Routledge.

Callon, M. (1999) The Role of Lay People in the Production and Dissemination of Scientific Knowledge. *Science, Technology and Society* 4(1):81–94.

Collins, H. (1993) The structure of Knowledge, *Social Research*, 60:95–116.

Cooper, M. (2008) *Life as Surplus: Biotechnology and Capitalism in the Neoliberal Era.* Seattle: University of Washington Press.

Dosi, G., 1988. The nature of the innovative process. In: Dosi, G. et al. (Eds.) Technical Change and Economic Theory. London: Pinter: pp. 221–238.

Dosi, G., Faille, M. and Marengo, L. (1993) Organisational Capabilities, Patterns of Knowledge Accumulation and Governance Structures in Business Firms: An Introduction. *Organisation Studies* 29: 1165–1185.

Gibson-Graham, J.K. (1996) *The End of Capitalism (As We Knew It): A feminist Critique of Political Economy.* Oxford UK and Cambridge USA: Blackwell Publishers.

Gibson-Graham, J. K. (2006) *A Postcapitalist Politics.* Minneapolis, MN: University of Minnesota Press.

Giddens, A. (1991) *Modernity and Self-Identity: Self and Society in the Late Modern Age.* CA: Stanford University Press.

Haraway, D. (1991) 'Situated knowledges: The science question in Feminism and the privilege of partial perspective', in Haraway, D. (ed.) *Simians, Cyborgs and Women: The Reinvention of Nature.* New York: Routledge.

Hardt, M. (1999) Affective Labour, *Boundary 2* 26(2): 89–100.

Hardt, M. and Negri, A. (2004) *Multitude: War and Democracy at the Age of Empire.* New York: The Penguin Press.

Ingold, T. (2000) *The perception of the environment: essays on livelihood, dwelling and skill.* London: Routledge.

Johnson, B. (1992) Institutional learning. In Lundvall, B. (Ed.), *National Systems of Innovation.* London: Pinter.

Knorr Cetina, K. (1981) *The Manufacture of Knowledge.* Oxford: Pergamon Press.

Lam, A. (2000) Tacit Knowledge, Organisational learning and Societal Institutions: An Integrated Framework, *Organization Studies* 21(3):487–513.
Latour, B. (1986) Visualisation and Cognition: Thinking with Eyes and Hands, *Knowledge and Society* 6:1–40.
Lazzarato, M. (1996) Immaterial Labor. In Hardt, M. and Virno, P. (eds.) *Radical Thought in Italy: A Potential Politics*, Minneapolis: University of Minnesota: pp. 133–50.
Morini, C. (2007) The Feminisation of Labour in Cognitive Capitalism, *Feminist Review* 87: 40–59.
Nonaka, I. (1994) A Dynamic Theory of Organisational Knowledge Creation, *Organisation Science* 5(1): 14–37.
Nonaka, I. and Takeuchi, H. (1995) *The Knowledge-Creating Company: How the Japanese Create the Dynamic Innovation*. New York: Oxford University Press.
Sunder Rajan, K. (2006) *Biocapital: The Constitution of Postgenomic Life*. Durham and London: Duke University Press.
Thacker, E. (2005) *The Global Genome: Biotechnology, Politics and Culture*. Cambridge, MA: MIT Press.
Wallace, H. (2010) Bioscience for Life? Who Decides what research is done in health and agriculture? Genewatch UK Report. March 2010.

References

AAI (2009) The Abuse of Supermarket Buyer Power in the EU Food Retail Sector: Preliminary Survey of Evidence, Myriam Vander Stichele, SOMO & Bob Young, Europe Economics on behalf of Agribusiness Accountability Initiative (AAI) Amsterdam.

Abramowitz, M. and David, P. (1996) Technological Change, Intangible Investments and Growth in the Knowledge-Based Economy: The US Historical Experience. In Foray, D. and Lundvall, B.A. (Eds.) *Employment and Growth in the Knowledge-based Economy*. Paris: OECD: pp. 35–60.

Agrawal, Arun (1995) 'Dismantling the Divide between Indigenous and Scientific Knowledge', *Development and Change* 26(3): 413–39.

Allen P., FitzSimmons M., Goodman M. and Warner K. (2003) Shifting plates in the agrifood landscape: the tectonics of alternative agrifood initiatives in California, *Journal of Rural Studies* 19(1): 61–75.

Allen, P., Van Dusen, D., Lundy, J. and Gliessman, S. (1991) Integrating social, environmental, and economic issues in sustainable agriculture, *American Journal of Alternative Agriculture* 6(1): 34–39.

Allen, P. and Wilson, A.B. (2008) Agri-food inequalities: globalization and localization, *Development* 51(4): 534–540.

Amin, A. and Cohedent, P. (2004) *Architectures of Knowledge: Firms, Capabilities and Communities*. Oxford: Oxford University Press.

Andersson, F., Samuelson, J., Larsen, K., Lagerkvist, C.J., Andersson, C., Bladm F., Samulesson, J. and Skargen, P. (2005) Farm Cooperation to Improve Sustainability, *Ambio* 34(4):38–87.

Antweiler, C. (1998) Local Knowledge and Local Knowing: An Anthropological analysis of contested "Cultural Products" in the Context of Development, *Anthropos* 93: 469–494.

Antweiler, C. (2004) Local Knowledge, Theory and Methods: an urban model from Indonesia. In Bicker, A., Sillitoe P. and Pottier, J. (eds.) Investigating Local Knowledge. New Directions, New Approaches. Ashgate: pp. 1–34.

Appadurai, A. (1996) *Modernity at large: Cultural dimensions of globalization.* Minneapolis: University of Minnesota Press.

Arrow, K. (1962) The Economic Implications of Learning by Doing. Review of Economic Studies, 29(3):155–173.

Asheim, BT and Coenen, L. (2006) Contextualising regional innovation systems in a globalising learning economy: on knowledge bases and institutional frameworks, *The Journal of Technology Transfer* 31:163–173.

Bager, T., Proost, J. (1997). Voluntary regulation and farmers' environmental behaviour in Denmark and The Netherlands, *Sociologia Ruralis* 37 (1): 79–98.

Baudrillard, J. (1972) *For a critique of the political economy of the sign.* St. Louis: Telos.

Bauen, A., Chambers, G., Houghton, M., Mirmolavi, B., Nair, S. Nattrass, L., Phelan, J. and Pragnell, M. (2016) Evidencing the Bioeconomy, Available onlineathttps://bbsrc.ukri.org/documents/1607-evidencing-the-bioeconomy-report/#:~:text=The%20bioeconomy%20is%20the%20production,transformative%20processes%20using%20biological%20resources.

Beck, U. (1992) *Risk society: towards a new modernity.* London: Sage.

Bell, J. (1997) *Doing Your Research Project.* Buckingham: Open University Press.

Benton, T. (1989) Marxism and natural limits: an ecological critique and reconstruction. *New Left Review*, 178:51–86.

Biodynamic Association (2011) About BDA. Available online at http://www.biodynamic.org.uk/.

Birch, K. (2007) Knowledge, Place and Power: Conceptualising Value Creation in Knowledge-Based Commodity Chains. Working paper. Centre for Public Policy for Regions. University of Glasgow.

Birch, K. (2009) The Knowledge-Space Dynamic in the UK Bioeconomy, *Area* 41(3): 273–284.

Birch, K, Levidow, L. and Papaioannou, T. (2010) Sustainable Capital? The Neoliberalisation of Nature in the European Knowledge-based Bio-economy, *Sustainability* 2: 2898–2918.

Blackler, F. (1995) Knowledge, Knowledge Work and Organizations: An Overview and Interpretation. *Organisation Studies*, 16(6): 1021–1046.

Blythman, J. (2004) *Shopped: The Shocking Power of British Supermarkets.* Fourth Estate.

Bourdieu, P. (1990) *The Logic of Practice.* Cambridge UK: Polity Press: Oxford UK: B. Blackwell.

Boyd, W, Prudham, S. and Schurman, R. (2001) Industrial dynamics and the problem of nature. *Society and Natural Resources*, 14:555–570.

Boyes, W. and Melvin, M. (2010) *Fundamentals of Economics*. South-Western: Cengage Learning.

Brennan, T. (2000) *Exhausting Modernity: Grounds for a new economy*. London and New York: Routledge.

Brinkley, I. (2006) *Defining the Knowledge Economy. Knowledge Economy Programme Report*. The Work Foundation. Available online at: http://www.flacso.edu.mx/openseminar/downloads/brinkley_S3.pdf.

Brinkley, I. and Lee, N. (2007) *The Knowledge economy in Europe. A Report prepared for the 2007 EU Spring Council*. The Work Foundation. Available online at: http://www.theworkfoundation.com/assets/docs/publications/80_Knowledge%20Economy%20EU%20Spring%20Council.pdf.

Brom, F.W.A. (2000) Food, consumer concerns, and trust: food ethics for a globalizing market, *Journal of Agricultural & Environmental Ethics* 12(2):127–139.

Brown, J.S. and Duguid, P. (1991) Organizational learning and communities of practice: towards a unified view of working, learning and innovation, *Organisation Science* 2(1):40–57.

Brown, S. and Getz, C. (2008) Towards domestic fair trade? Farm labor, food localism, and the 'family scale' farm, *GeoJournal* 739(1):11–22.

Brubaker, R. (1984) *The Limits of Rationality. An Essay on the Social and Moral Thought of Max Weber*. London, George Allen & Unwin.

Bryant, R. 2001. Political Ecology: A Critical Agenda for Change? In Castree, N. and Braun, B. (eds.) *Social Nature: Theory, Practice and Politics* Oxford: Blackwell: pp. 151–169.

Bruckmeier, K. and Tovey, H. (2008) Knowledge in sustainable rural development: from forms of knowledge to knowledge processes, *Sociologia Ruralis* 48(3): 313–329.

Burch, D. and Lawrence, G. (2007) *Supermarkets and agri-food supply chains: transformations in the production and consumption of foods*. Edward Elgar Publishing.

Burton-Jones, A. (1999) *Knowledge Capitalism. Business, Work and Learning in the New Economy*. Oxford: Oxford University Press.

Buscher, M., Urry, J. and Witcger, K. (2011) *Mobile Methods*. London: Routledge Academic Publishers.

Buttel, F. (1997). Some Observations on Agro-Food Change and the Future of Agricultural Sustainability Movements. In Goodman, D. and Watts, M. (Eds.) *Globalising Food: Agrarian Questions and Global Restructuring*. London: Routledge: pp. 344–365.

Buttel, F. (2005) Ever Since High Tower: The Politics of Agricultural Research Activism in the Molecular Age, *Agriculture and Human Values* 22(3):275–283.

Callon, M. (1999) The Role of Lay People in the Production and Dissemination of Scientific Knowledge. *Science, Technology and Society* 4(1):81–94.

Callon, M., Lascoumes, P. and Barthe, Y. (2009) *Acting in an uncertain world: An essay on technical democracy*. Translated from French by Graham Burchell. Cambridge, MA, USA, London: MIT Press.

Campbell, A. (1984) *The Girls in the Gang*. Oxford: Basil Blackwell.

Castells, M, (1996). *The Rise of the Network Society*. Oxford: Blackwell.

Castoriadis, C. (1987) *The imaginary institution of society*. Translated by Kathleen Blamey. Cambridge, MA: MIT Press.

Clarke, N., Cloke, P. Barnett, C. and Malpass, A. (2008) The spaces and ethics of organic food, *Journal of Rural Studies* 24:219–230.

Collier, A. (2003) *In Defense of Objectivity*, London: Routledge.

Collins, H. (1993) The structure of Knowledge, *Social Research*, 60:95–116.

Collins, H.M. and Evans, R. (2002) The Third Wave of Science Studies: Studies of Expertise and Experience, *Social Studies of Science* 32(2):235–296.

Cook, I. and Crang, M. (1995), *Doing Ethnographies—Concepts and Techniques in Modern Geography 58*. Norwich: Environmental Publications.

Cooper, M. (2007) Life, autopoiesis, debt: inventing the bioeconomy, *Distinktion* 14: 25–43.

Cooper, M. (2008) *Life as Surplus: Biotechnology and Capitalism in the Neoliberal Era*. Seattle: University of Washington Press.

Cowley, J. Property Scandal. *New Statesman*. 20 September 2004. Available online at http://www.newstatesman.com/200409200005.

CPRE (2006) The Real Choice: how local foods can survive the supermarket onslaught. Campaign for Protection of Rural England. Available online at www.cpre.org.uk.

Crang, M. (1997) Analyzing Qualitative Materials. In Flowerdew, R. and D. Martin (eds.) *Methods in Human Geography*, Essex: Pearson Education Limited: pp. 183–196.

Crang, M. (2001) Filed work: making sense of group interviews. In Limb, M. and Dwyer, C., (eds.) *Qualitative Methodologies for Geographers*. London: Arnold: pp. 215–233.

Cumbria Fells and Dales Local Action Group (2008) Development Strategy, Rural Development Programme for England, www.fellsanddales.org.uk. Available online at http://www.cumbria.gov.uk/business/rdpe/fellsanddales/fellsanddales.asp.

Daston, L. and Galison, P. (1992) The Image of Objectivity. *Representations* 40: 81–128.

Davies, C., (1999) *Reflexive Ethnography*. London: Routledge.

Davis, J. and Hinshaw, K. (1957) *Farmer in a Business Suit*. New York: Simon and Schuster, Inc.

Deflem, D. (2003) The sociology of the sociology of money: Simmel and the contemporary battle of the classics. *Journal of Classical Sociology* 31(1): 67–96.

DEFRA (2010) Food Statistics Pocket Book. Available online at www.defra.gov.uk.

Degen, M.M. (2010) The urban green: passionate involvements with urban natures. In Miles, M. and Degen, M. (eds.) *Culture and Agency: Contemporary Culture and Urban Change.* University of Plymouth Press: pp. 58–75.

Derrida, J. (2001) History of the Lie: Prolegomena. In Rand, R. (ed.) *Futures of Jacques Derrida.* Stanford: Stanford University Press: pp. 65–98.

Denzin, N. and Lincoln, Y. (2000) *Handbook of Qualitative Research.* London: Sage Publications.

DG Research (2005) New Perspectives on the Knowledge-Based Bio-Economy: Conference Report (Brussels: DG Research). Available at: http://ec.europa.eu/research/conferences/2005/kbb/report_en.html.

DG Research (2006) Framework Programme 7, Theme 2: Food, Agriculture, Fisheries and Biotechnology (FAFB), 2007 Work Programme, Commission of the European Communities: Brussels, Belgium, 2006.

DG Research (2007) FP7 'Presentation on KBBE' (Brussels: DG Research), Available at: ftp://ftp.cordis.europa.eu/pub/fp7/kbbe/docs/about-kbbe.pdf. Retrieved October 15, 2009.

Dosi, G., 1988. The nature of the innovative process. In: Dosi, G. et al. (Eds.) Technical Change and Economic Theory. London: Pinter: pp. 221–238.

Dosi, G., Faille, M. and Marengo, L. (1993) Organisational Capabilities, Patterns of Knowledge Accumulation and Governance Structures in Business Firms: An Introduction. *Organisation Studies* 29: 1165–1185.

Dosi, G., Faille, M. and Marengo, L. (2008) Organisational Capabilities, patterns of knowledge accumulation and Governance Structures in Business Firms: An Introduction. *Organisational Studies* 29: 1165–1185.

Dretske, F. (1981), *Knowledge and the Flow of Information.* Cambridge, MA: MIT Press.

Dreyfus, H. L. and Rabinow, P. (1982) Michel Foucault: beyond Structuralism and Hermeneutics. University of Chicago.

Drucker, P.F. (1993) *Post-Capitalist Society.* New York: Harper Collins.

Drucker, P.F. (2002) *Managing in the Next Society.* New York: St. Martin's Griffin.

Du Puis, E.M. and Goodman, D. (2005) Should we got home to eat? Towards a reflexive politics of localism, *Journal of Rural Studies* 21:359–71.

Durbin, S. (2006) Who Gets to be a Knowledge Worker? The case of UK call centres. In Walby, S., Gottfried, H., Gottschall, K. & Osawa, M. (eds.) *Gendering the Knowledge Economy: Comparative Perspectives.* Palgrave Macmillan: pp. 228–249.

Eaves, Y. (2001) A synthesis technique for grounded theory data analysis, *Journal of Advanced Nursing* 35(5):654–663.

Eden, S., Bear, C. and Walker, G. (2008) Mucky Carrots and Other Proxies: problematising the knowledge-fix for sustainable and ethical consumption, *Geoforum* 39(2):1044–1057.

EC (1993) Growth, competitiveness, employment: the challenges and ways forward into the 21st century—White Paper, Parts A and B. COM (93) 700 final/A and B, 5 December 1993, Bulletin of the European Communities, Supplement 6/93.

EC (2000) *The Lisbon European Council—An Agenda of Economic and Social Renewal for Europe*. Brussels: The European Commission, DOC/00/7.

EC (2002) Life Sciences and Biotechnology: A Strategy for Europe. COM(2002) 7. Available at: http://ec.europa.eu/biotechnology/pdf/com2002-27_en.pdf.

EC (2007) *En route to the Knowledge-based Bioeconomy*. 'Cologne Paper'. Conference and Workshop Findings. Available at http://www.bio-economy.net/reports/files/koln_paper.pdf.

Epstein, S. (2007) *Inclusion: The Politics of Difference in Medical Research*. Chicago: University of Chicago Press.

Evans-Pritchard, E.E. (1967) *Introduction to The Gift, by Marcel Mauss*. New York: Norton.

European Commission (2008) Growth and Jobs: Background (Brussels: European Commission), Available online at: http://ec.europa.eu/growthandjobs/faqs/background/index_en.htm#bg01.

FAAN (2010) Local food systems in Europe: case studies from five countries and what they imply for policy and practice. Graz, IFZ. Available online at http://www.faanweb.eu/sites/faanweb.eu/files/FAAN_Booklet_PRINT.pdf.

FARMA. (2006) Farmers' markets in the UK: Nine years and counting. Southampton, Available online at www.farma.org.uk. Retrieved January 20, 2010.

Feagan, R. (2007) The place of food: Mapping out the 'local' in local food systems, *Progress in Human Geography* 31(1):23–42.

Feagan, R., Morris, D. and Krug, K. (2004) Niagara Region Farmers' Markets: local food systems and sustainability considerations. *Local Environment* 9(3):235–254.

Feenstra, G. (1997) Local food systems and sustainable communities *American Journal of Alternative Agriculture* 12:28–36.

Fielding, N. (1988) *Joining Forces: Police training, Socialization and Occupational Competence*. London: Routledge.

Fine, B. and Leopold, I. (1994) *The World of Consumption*. London: Routledge.

Flick, U. (2006) *An Introduction to Qualitative Research*. London: Sage Publications.

Fonte, M. (2008) Knowledge, Food and Place. A Way of Producing, a way of knowing, *Sociologia Ruralis* 48(3): 200–222.

Fonte, M. and Grando, S. (2006) A Local Habitation and a Name: Local Food and Knowledge Dynamics in Sustainable Rural Development, *CORASON project*. Available online at www.corason.hu.

Fonte, M. and Papadopoulos, A. (2010) *Naming Food After Places: Food Relocalisation and Knowledge Dynamics in Rural Development*. Ashgate.

Food and Agriculture Organisation (2003) *The State of Food Insecurity in the World 2003*. Rome, FAO. Available online at ftp://ftp.fao.org/docrep/fao/006/j0083e/j0083e00.pdf.

Food Futures. (2007) *A Food Strategy for Manchester.* Available online at http://www.foodfutures.info/site/images/stories/food%20futures%20strategy%202007.pdf.
Foray, D. and Lundvall, B-A. (1996) *Employment and Growth in the Knowledge-Based Economy.* Paris: OECD.
Franklin, A. (2002) *Nature and Social Theory.* London: Sage.
Friedrichs, J. and Ludtke, H. (1975) *Participant Observation: theory and practice.* Saxon House and Lexington Books.
Fuller, S. (2006) Introduction. In Stehr, N, Henning, C and B. Weiler (Eds.) *The Moralization of the Markets.* New Brunswick, Transaction Publishers.
Funtowicz, S.O. and Ravetz, J.R. (1994a) Uncertainty, Complexity and Post-Normal Science. *Environmental Toxicology and Chemistry* 13(12): 1981–1984.
Funtowicz, S.O. and Ravetz, J.R. (1994b) Emergent Complex Systems. *Futures* 26(6): 568–582.
Getz, C., Brown, S. and Shreck, A. (2008) Class Politics and Agricultural Exceptionalism in California's Organic Agriculture Movement. *Politics Society* 36(4): 478–507.
Gibb, R (2005) *Greater Manchester: A panorama of people and places in Manchester and its surrounding towns.* Myriad, 13.
Gibbons, M., C. Limoges, H. Nowotny, S. Schwartzmann, P. Scott and M. Trow (1994) *The new production of knowledge: the dynamics of science and research in contemporary societies.* London: Sage.
Gibson-Graham, J.K. (1996) *The End of Capitalism (As We Knew It): A feminist Critique of Political Economy.* Oxford UK and Cambridge USA: Blackwell Publishers.
Gibson-Graham, J. K. (2006) *A Postcapitalist Politics.* Minneapolis, MN: University of Minnesota Press.
Giddens, A. (1976) *New rules of sociological method.* London: Hutchinson.
Giddens, A. (1990) *The Consequences of Modernity.* Cambridge: Polity Press.
Giddens, A. (1991) *Modernity and Self-Identity: Self and Society in the Late Modern Age.* CA: Stanford University Press.
Gilbert, E. and Helleiner, E. (1999) Introduction—Nation-states and money: historical contexts, interdisciplinary perspectives. In E. Gilbert and E. Helleiner (eds.) *Nation-states and Money: The past, present and future of national currencies.* London: Routledge: pp. 1–21.
Godin, B. (2006) The Knowledge-Based Economy: Conceptual Framework or Buzzword?, *The Journal of Technology Transfer* 31(1): 17–30.
Gold, R. (1969) Roles in sociological field observation, in G. McCall and J. Simmons (eds.) *Issues in participant Observation: A text and Reader.* London: Addison Wesley.
Goldman, M. (2004) Imperial science, imperial nature: environmental knowledge for the World (Bank) in Jasanoff, S. and Martello, M.L. (eds.) *Earthly politics:*

local and global in environmental governance. Cambridge, MA: MIT Press: pp. 55–80.

Gonzalez-Velez, M. (2002) Assessing the Conceptual Use of Social imagination in Media Research *Journal of Communication Inquiry*, 26: 349–353.

Goodman, D. (2003) The quality 'turn' and alternative food practices: reflections and agenda. *Journal of Rural Studies* 19:1–7.

Goodman, D. and Goodman, M. (2007) Localism, Livelihoods and the 'Post-Organic': Changing Perspectives on Alternative Food Networks in the United States. In Maye, D, Holloway, L. and Kneafsy, M. (eds.) *Alternative Food Geographies Representation and Practice.* Elsevier.

Goodman, D. and Redclift, M. (1991) *Refashioning Nature.* London: Routledge.

Goodman, D. and Watts, M. J. (1997) *Globalising food. Agrarian Questions and Global Restructuring.* London and New York: Routledge.

Goodman, D., Sorj, B. and Wilkinson, J. (1987) *From farming to biotechnology.* Oxford: Basil Blackwell.

Goodman, M.K., Maye, D. and Holloway, L. (2010) Ethical foodscapes?: Premises, promises, and possibilities, *Environment and Planning A* 42(8): 1782–1796.

Gottlieb, R. (2001) Environmentalism Unbound: Exploring new pathways for change. Cambridge, MA: The MIT Press.

Gottweis, H. (1998) *Governing Molecules: The Discursive Politics of Genetic Engineering in Europe and the United States.* MIT Press.

Gouldner, A.W. (1960) The Norm of Reciprocity: A Preliminary Statement, *American Sociological Review* 25: 161–178.

Graeber, D. (2001) *Toward an Anthropological Theory of Value.* Palgrave Macmillan.

Gray, J. (1999) *Falls Down-The Delusion of Global Capitalism.* London: Granta Books.

Grigg, D. (1982) *The World Food Problem, 1950–80.* Oxford: Basil Blackwell.

Gronow, J. and Ward, A. (Eds.) (2001) *Ordinary Consumption.* London: Routledge.

Guthman, J. (2004) The trouble with 'organic lite' in California: a rejoinder to the 'conventionalisation' debate, *Sociologia Ruralis* 44(3): 301–316.

Hammersley, M. (1991) Some reflections on ethnography and validity, *Qualitative Studies in Education* 5(3):195–203.

Hammersley, M. (1992) *What's Wrong with Ethnography?* London: Routledge.

Haraway, D. (1991) 'Situated knowledges: The science question in Feminism and the privilege of partial perspective', in Haraway, D. (ed.) *Simians, Cyborgs and Women: The Reinvention of Nature.* New York: Routledge.

Hardt, M. (1999) Affective Labour, *Boundary 2* 26(2): 89–100.

Hardt, M. and Negri, A. (2004) *Multitude: War and Democracy at the Age of Empire.* New York: The Penguin Press.

Hartwick, E. (1998) Geographies of consumption: a commodity-chain approach, *Environment and Planning D: Society and Space* 16: 423–437.

Hassanein, N. (2003) Practicing Food democracy: a pragmatic politics of transformation, *Journal of Rural Studies* 19:77–86.

Havelock, R.G. (1969) *Planning for Innovation: A Comparative Study of the literature on the dissemination and utilisation of scientific knowledge.* Ann Arbor, University of Social Research, University of Michigan: p. 533.

Hayek, F. (1945) The use of Knowledge in Society, *American Economic Review* 35(4):519–30.

Hayek, F. (1976) *Choice in Currency: A Way to Stop Inflation.* London: Institute of Economic Affairs.

Hedges, A. and Zykes, W. (2003) *Local food: a report on qualitative research.* London: Food Standards Agency. Available online at http://www.food.gov.uk/multimedia/pdfs/localqualitative.pdf. Retrieved on January 20, 2011.

Heelas, P. (1996) Introduction: Detraditionalisation and Its Rivals. In Heelas, P., Lash, S. and Morris, P. (Eds.) *Detraditionalisation: Critical Reflections on Authority and Identity.* Cambridge: Blackwell Publishers.

Henry, R. and Pollard, J. (2000) Introduction: Capitalising on Knowledge, *Geoforum* 31: v–vii.

Henderson, E. (1998) Rebuilding local food systems from the grassroots up, *The Monthly Review: An Independent Socialist Magazine* 50:112–124.

Henderson, E. (2000) Rebuilding local food systems from the grassroots up. In Magdoff, F., Bellamy Foster, J. and Buttel, F.H. (Eds.) *Hungry for Profit: The agribusiness Threat to Farmers, Food and the Environment.* New York: Monthly Review Press: pp. 175–188.

Hendrickson, M. and Hefferman, W. (2002) Opening spaces through relocalization: locating potential resistance in the weaknesses of the global food system, *Sociologia Ruralis* 42:347–69.

Heyl, B. (2001) Ethnographic Interviewing. In P. Atkinson, S. Coffey, and Delamont, S. (Eds.) *Handbook of Ethnography.* SAGE: pp. 369–383.

Hinchliffe, S., Degen, M., Kearnes, M. and Whatmore, S. (2005) Urban wild things: a cosmopolitical experiment, *Environment and Planning D: Society and Space* 23 (5): 643–658.

Hinrichs, C. (2000) Embeddedness and local food systems: notes on two types of direct agricultural market, *Journal of Rural Studies* 16: 295–303.

Hinrichs, C. (2003) The practice and politics of food system localization, *Journal of Rural Studies* 16:33–45.

Hodgson, G. (1999) *Economics and Utopia: Why the Learning Economy is not the End of History.* London: Routledge.

Horton, D. (2003) Green distinctions: the performance of identity among environmental activists. In Szerszynski, B., Heim, W. and Waterton, C. (eds.) *Nature performed: Environment, Culture and performance.* Cambridge: Blackwell: pp. 63–77.

Humphreys, L. (1970) *Tea Room Trade.* London: Duckworth.

Huxley, R. (2003) *A Review of the UK Food Market*. Report for Cornwall Agricultural Council and Taste of West. Available online at http://www.objectiveone.com/ob1/pdfs/uk_food_market_review.pdf.

Ikerd, J. (1993) Two related but distinctly different concepts: organic farming and sustainable agriculture, *Small Farm Today* 10(1):30–31.

Ilbery, B. and Bowler, I. (1998) From agricultural productivism to post-productivism. In B. Ilbery (ed.) *The Geography of Rural Change*. London: Longman.

Ilbery, B. and Kneafsy, M. (1999) Niche Markets and regional speciality food products in Europe: towards a research agenda, *Environment and Planning A* 31:2207–2222.

Ilbery, B. and Maye, D. (2005) Alternative (shorter) food supply chains and specialist livestock products in Scottish-English borders, *Environment and Planning A* 37(4):823–844.

Ilbery, B., Watts, D., Simpson, S., Gilg, A. and Little, J. (2006) Mapping local foods: evidence from two English Regions, *British Food Journal* 108(3):213–225.

Iles, A. (2005) Learning in sustainable agriculture: food miles and missing objects, *Environmental Values* 14(2): 63–183.

Ingold, T. (2000) *The perception of the environment: essays on livelihood, dwelling and skill*. London: Routledge.

Ingold, T. and Kurtilla, T. (2000) Perceiving the environment in Finnish Lapland. *Body and Society*, 6(3/4):183–196.

Jackson, P. (2001) Making sense of qualitative data. In Limb, M. and Dwyer, C. (eds.) *Qualitative Methodologies for Geographers*, London: Arnold: pp. 199–214.

Jackson, P, Russel, P. and Ward, N. (2006) Mobilising the commodity chain concept in the politics of food and farming, *Journal of Rural studies* 22: 129–141.

Jackson, P. Ward, N. and Russel, P. (2009) Moral economies of food and geographies of responsibility, *Transactions of the Institute of British Geographers* 34:12–24.

James, W. (1950) *The Principles of Psychology*. New York: Dover.

Jasanoff, S. and Martello, M.L. (2004) Conclusion: knowledge and governance. In S. Jasanoff and M.L. Martello (eds.), *Earthly politics: local and global in environmental governance* Cambridge, MA: Massachusetts Institute of Technology. pp. 335–350.

Jasanoff, S., Kim, S-H. and Sperling, St. (2008) Sociotechnical Imaginaries and Science and Technology Policy: A Cross-National Comparison. Project Summary.

Jessop, B. (2005) Cultural Political Economy, the Knowledge-based Economy and the State. In Barry, A. and Slater, D. (eds.) The technological Economy. Routledge. pp. 142–164.

Jessop, B. (2007) Knowledge as a fictitious commodity : insights and limits of a Polanyian perspective. In: Bugra, Ayse and Agartan, Kaan, (eds.) *Reading Karl Polanyi for the twenty-first century: market economy as political project*. Palgrave, Basingstoke.

Jessop, B. (2008) *State power: a strategic-relational approach*. Cambridge: Polity Press.
Johnson, B. (1992) Institutional learning. In Lundvall, B. (Ed.), *National Systems of Innovation*. London: Pinter.
Juma, C. and Konde, V. (2001) The new bioeconomy: Industrial and environmental biotechnology in developing countries. UNCTAD/DITC/TED/12.
Kirwan, J. (2004) Alternative Strategies in the UK Agro-food system: Interrogating the Alterity of Farmers' Markets, *Sociologia Ruralis* 44(4):395–415.
Kloppenburg, J. (1988) *First The Seed: The Political Economy of Plant Biotechnology, 1492–2000*. Cambridge University Press.
Kloppenburg, J. Jr (1991) Social theory and the de/reconstruction of agricultural science: local knowledge for an alternative agriculture, *Rural Sociology* 56 (4): 519–548.
Kloppenburg, J., Henrickson, J. and Stevenson, G.W. (1996) Coming in to the foodshed, *Agriculture and Human Values* 13:33–42.
Kloppenburg, J., Lezberg, S., De Master, K., Stevenson, G.W. and Hendrickson, J. (2000) Tasting Food, tasting sustainability: Defining the Attributes of an alternative food system with competent, ordinary people, *Human Organisation* 59 (2): 177–186.
Knickel, K., Susanne von Münchhausen, Henk Renting and Sarah Peter (2008) Supporting collective action in alternative food networks: Findings from 18 in-depth case studies in ten European countries, Second International Working Conference for Social Scientists "Sustainable Consumption and Alternative Agri-Food Systems", 27–30 May 2008, Arlon. Available online at http://www.suscons.ulg.ac.be. Retrieved March 17, 2009.
Knorr Cetina, K. (1981) *The Manufacture of Knowledge*. Oxford: Pergamon Press.
Kuhn, T. (1962) *The Structure of Scientific Revolutions*. University of Chicago Press.
Via Campesina (1996) *Tlaxcala Declaration of the Via Campesina*. April 1996. Available online at http://www.viacampesina.org/en/index.php?option=com_content&view=article&id=445:ii-international-conference-of-the-via-campesina-tlaxcala-mexico-april-18-21&catid=32:2-tlaxcala-1996&Itemid=48.
Lacroix, A. (1981) *Tranformations du process de travail agricole; indicidences de 'industrialisation sur les conditions de travail paysannes*. Grenoble: Institute National de la Recherche Agronomique.
Lacy, W. (2000) Empowering Communities through public work, science and local food systems: revisiting democracy and globalisation, *Rural Sociology* 65:3–26.
Lakoff, G. and Johnson, M. (1980) *Metaphors We Live By*. Chicago, IL: University of Chicago Press.
Lam, A. (2000) Tacit Knowledge, Organisational learning and Societal Institutions: An Integrated Framework, *Organization Studies* 21(3):487–513.

Lang, T. (1999) Food policy for the 21st century: can it be both radical and reasonable? In Koc, M., MacRae, R. Mougeot, L.J.A., Welsh, J. (Eds.) *For Hunger-proof Cities: Sustainable Urban Food Systems*. Otawa, International Development Research Centre.

Lang, T. (2009) Re-shaping the food system for ecological public health, *Journal of Hunger and Environmental Nutrition* 4(3/4): 315–335.

Lang, T. and Heasman, M. (2004) *Food Wars: Public health and the battle for mouths minds and markets*. London: Earthscan.

Lang, T., Barling, D. and Caraher, M. (2009) *Food Policy: Integrating Health, Environment and Society*. Oxford: Oxford University Press.

Latour, B. (1986) Visualisation and Cognition: Thinking with Eyes and Hands, *Knowledge and Society* 6:1–40.

Latour, B. (1987) *Science in action*. Cambridge, MA: Harvard University Press.

Latour, B. (1993) *We Have Never Been Modern*. Cambridge, MA: Harvard University Press.

Law, J. 2004. STS a Method. Lancaster University. Available online at http://heterogeneities.net/publications/Law2015STSAsMethod.pdf.

Lazzarato, M. (1996) Immaterial Labor. In Hardt, M. and Virno, P. (eds.) *Radical Thought in Italy: A Potential Politics*, Minneapolis: University of Minnesota: pp. 133–50.

Lee, R. (2000) Shelter from the storm? Geographies of regard in the worlds of horticultural consumption and production, *Geoforum* 31:137–157.

Levenstein, H. (1993) The Paradox of Plenty: A Social History of Eating in Latin America. Oxford: Oxford University Press.

Levidow, L. (2008) European quality agriculture as an alternative bio-economy. In Guido Ruivenkamp, Shuji Hisano and Joost Jongerden (Eds.) *Reconstructing Biotechnologies: Critical Social Analyses*. Wageningen Academic: pp. 185–205.

Levidow, L. and Boschert, K. (2008) Coexistence or contradiction? GM crops versus alternative agricultures in Europe, *Geoforum* 39(1): 174–190.

Levidow, L. and Carr, S. (2007) GM crops on trial: technological development as a real-world experiment, *Futures* 39(4): 408–431.

Levidow, L. and Psarikidou, K. (2011) Food Relocalisation for Environmental Sustainability in Cumbria, *Sustainability* 2(1):692–719.

Levidow, L. and Psarikidou, K. (2012) Making Local Food Sustainable in Manchester. In Viljoen, A. and Wiskerke, J.S.C. (eds.) *Sustainable Food Planning: Evolving Theory and Practice*. Wageningen: Wageningen Academic Publishers: pp. 207–220.

Levidow, L., Birch, K., Papaioannou, T. 2012a. Divergent paradigms of European Agro-Food Innovation: the Knowledge-based Bioeconomy (KBBE) as an R&D agenda, *Science, Technology and Human Values*, 38(1): 94–125.

Levidow, L., Birch, K. and Papaioannou, T. 2012b. EU agri-innovation policy: two contending visions of the bio-economy, *Critical Policy Studies* 6(1): 40–65.

Levidow, L., Carr, S. and Wield, D. (2005) EU regulation of agri-biotechnology: precautionary links between science, expertise and policy, *Science & Public Policy* 32(4): 261–276.

Levidow, L., Price, B., Psarikidou, K., Szerszynski, B. and Wallace, H. (2010) Urban Agriculture as Community Engagement, *Urban Agriculture Magazine* 24:43–45.

Lenin, V. I. (1938) *Theory of the Agrarian Question*. New York, NY: International Publishers, V. I. Lenin, Selected Works, Vol. XII.

Lewis, M. (1999) The new new thing: A Silicon Valley Story. New York: W.W.Norton.

Little, R., Maye, D. and Ilbery, B. (2010) Collective purchase: moving local and organic foods beyond the niche market, *Environment and Planning A* 42(8):1797–1813.

Lofland, J. and Lofland, L. (1984) *Analysing Social Settings: A Guide to Qualitative Observation and Analysis*, 2nd edn. Belmont, CA: Wadsworth.

Lorwin, L. L. (1931) Exploitation. In E. Seligman and A. Jonson (Eds.) *The Encyclopaedia of the Social Sciences*. Vol. 6. New York.

Lowe, P., Murdoch, J., Marsden, T., Munton, R., Flynn, A. (1993) Regulating the new rural spaces: the uneven development of land, *Journal of Rural Studies* 9:205–222.

Lundvall, B. and Johnson, B. (1994) The learning economy, *Journal of Industry Studies* 1(2): 23–42.

Machlup, F. (1980). *Knowledge and knowledge production. Knowledge: its creation, distribution, and economic significance*. Vol. I. Princeton, NJ: Princeton University Press.

Maffesoli, M. (1993a) Introduction, *Current Sociology* 41(2): 60–67.

Maffesoli, M. (1993b) The imaginary and the sacred in Durkheim's sociology, *Current Sociology* 41 (2): 1–5.

Manchester City Council. (2011a) *Indices of Multiple Deprivation 2010. Analysis for Manchester*. Available online at http://www.manchester.gov.uk/downloads/download/414/research_and_intelligence_population_publications_deprivation.

Manchester City Council. (2011b) *Manchester: A City for Everyone. Promoting Equality and Inclusion*. http://www.manchester.gov.uk/download/210/a_city_for_everyone.

Marsden, T.K. (1998) New rural territories: Regulating the differentiated rural spaces, *Journal of Rural Studies* 14(1):107–117.

Marsden, T.K. (2004) The quest for ecological modernisation: re-spacing rural development and agri-food studies, *Sociologia Ruralis* 44:129–146.

Marsden, T.K. and Smith, E. (2005) Ecological entrepreneurship: sustainable development in local communities through quality food production and local branding, *Geoforum* 36:440–451.

Marsden, T.K. and Sonnino, R. (2008) Rural development and the regional state: Denying multifunctional agriculture in the UK, *Journal of Rural Studies* 24:422–431.

Marsden, T. and Farioli, F. (2015) Natural Powers: From the Bio-economy to the Eco-economy and sustainable Place-making, *Sustainability Science* 10: 331–344.

Marsden, T., Murdoch, J. and Morgan, K. (1999) Sustainable agriculture, food supply chains and regional development: editorial introduction, *International Planning Studies* 4(3):295–301.

Marsden, T.K., Banks, J. and Bristow, G. (2000) Food supply chain approaches: exploring their role in rural development, Sociologia Ruralis 40:424–438.

Marx, K. (1956) *Capital*, vol. II. London: Lawrence and Wishart.

Marx, K. (1973) *Grundisse*. New York, NY: Vintage Books.

Marx, K. (1976) Capital: A Critique of Political Economy, Vol. 1. Harmondsworth: Penguin.

Marx, K. (1978) *The Marx-Engels Reader*. Edited by R.C. Tucker. New York: W.W. Norton.

Marx, K. and Engels, F. (1987) *Collected Works*, vol. 3. London.

Mauss, M. (1967) *The Gift: forms and functions of exchange in archaic societies*. London: Routledge.

May, T. (1991) *Probation: Politics, Policy and Practice*. Buckingham: Open University Press.

May, T. (2001) *Social Research: Issues, methods and process*. Buckingham, Philadelphia: Open University Press.

McMichael, P. (2000) The power of food. Agriculture and Human Values 17:21–33.

McMichael, P. (2009) A Food Regime Genealogy, *The Journal of Peasant Studies* 36(1): 139–169.

Mendras, H. (1970) *The Vanishing Peasant: Innovation and Change in French Agriculture*, Cambridge: Cambridge University Press.

Menger, C. (1976) *Principles of Economics*. New York: New York University Press.

Merton, R.K. (1942) *The Normative Structure of Science* In: Merton, Robert King (1973). *The Sociology of Science: Theoretical and Empirical Investigations*. Chicago: University of Chicago Press.

Merton, R.K. (1948) The Self-Fulfilling Prophecy, *The Antioch Review* 8(2): 193–210.

Mintel (2001) *Regional Eating and Drinking Habits*. Market Intelligence. December 2001. London: Mintel International Ltd.

Moore, O. (2004) What Farmers' Markets Say about the Post-organic Movement in Ireland. In Holt, G.C. and Reed, M. (eds.) *Sociological perspectives of Organic Agriculture*. CAB International.

Morgan, K. and Murdoch, J. (2000) Organic versus Conventional Agriculture: Knowledge, Power and Innovation in the Food Chain, *Geoforum* 31:159–173.

Morgan, K. J. and Sonnino, R. (2007) Empowering Consumers: The Creative Procurement of School Meals in Italy and the UK, *International Journal of Consumer Studies* 31(1):19–25.

Morgan, K., Marsden, T. and Murdoch, J. (2006) *Worlds of food: place, power, and provenance in the food chain* Oxford: Oxford University Press.
Morgan, W. (1993) The Wine appellation as Territory in France and California, *Annals of the Association of the American Geographers* 83(4): 694–717.
Morini, C. (2007) The Feminisation of Labour in Cognitive Capitalism, *Feminist Review* 87: 40–59.
Morris, C. (2006) Negotiating the boundary between state-led and farmer approaches to knowing nature: an analysis of UK agri-environment schemes, *Geoforum* 37(1):113–127.
Morris, C. and Buller, H. (2003) The local food sector: A preliminary assessment of its form and impact in Gloucestershire, *British Food Journal* 105(8):559–566.
Morris, C. and Young, C. (2000) 'Seed to shelf', 'teat to table', 'barley to beer' and 'womb to tomb': discourses of food quality and quality assurance schemes in the UK, *Journal of Rural Studies* 16:103–115.
Murdoch, J., Marsden, T. and Banks, J. (2000) Quality, nature and embeddedness: some theoretical considerations in the context of the food sector, *Economic Geography* 76(2):107–125.
Murdoch, J. and Miele, M. (2004) A new aesthetic of food? Relational reflexivity in the 'alternative' food movement. In M. Harvey, A. McMeekin and A. Warde (Eds.) *Qualities of Food: New Dynamics of Innovation and Competition.* Manchester: Manchester University Press: pp. 156–175.
Natural England (2008) *State of the Natural Environment 2008*. Available online at www.naturalengland.org.uk. Retrieved on April 2, 2011.
Newby, H. (1987) *Country Life: A Social History of Rural England.* London: Weidenfeld and Nicolson.
Newby, H., Bell, C, Rose, D. and Saunders, P. (1978) *Property, Paternalism and Power: Class and Control in Rural England.* London, Hutchinson & Co.
Nisikawa, M. and Tanaka, K. (2007) Are Care-Workers Knowledge-Workers? In Walby, S., Gottfried, H., Gottschall, K. and Osawa, M. (eds.) *Gendering the Knowledge Economy: Comparative Perspectives.* Palgrave Macmillan: pp. 207–227.
Nonaka, I. (1994) A Dynamic Theory of Organisational Knowledge Creation, *Organisation Science* 5(1):14–37.
Nonaka, I. and Takeuchi, H. (1995) *The Knowledge-Creating Company: How the Japanese Create the Dynamic Innovation.* New York: Oxford University Press.
North, P. (2007) *Money and Liberation: the micro-politics of alternative currency movements.* Minneapolis: University of Minnesota Press.
Nowotny, H, Scott, P. and Gibbons, M. (2001) *Re-thinking science. Knowledge and the public in an age of uncertainty.* Cambridge: Polity Press.
Nussbaum, M.C. (2001) *Upheavals of Thought: The Intelligence of Emotion.* Cambridge: Cambridge University Press.
NWDA (2008) More resources to grow rural economy, *315 Magazine*, Issue 15, June, 2008. Available online at http://www.nwda.co.uk/news%2D%2Devents/features/building-communities/resources-grow-rural-economy.aspx. Retrieved on July 20, 2008.

Nygren, A. (1999) Local Knowledge in the Environment_Development Discourse: From dichotomies to situated knowledges, *Critique of Anthropology* 19:267–288.

OECD (1996) *The Knowledge-based Economy*. Paris: OECD.

OECD (2003), *Harnessing Markets for Biodiversity: Towards Conservation and Sustainable Use*. Paris.

OECD (2005) *The Bioeconomy to 2030: Designing a Policy Agenda. Scoping Paper*. OECD Headquarters. 2 November 2005.

OECD (2006) *The Bioeconomy to 2030: Designing a Policy Agenda*. Scoping Paper. Informal Experts' Meeting. OECD Paris, 6 March 2006.

OECD (2012) Gross Value Added. Glossary of Statistical Terms. OECD. Available Online at http://stats.oecd.org/glossary/detail.asp?ID=1184. Retrieved on 25 March 2012.

ONS (2005) The UK's major urban areas Office for National Statistics—based on 2001 census figures.

Pardey, P.G. and Beintema, N.M. (2001) Slow Magic: Agricultural R&D A Century After Mendel. Agricultural Science and Technology Indicators Initiative. International Food Policy Research Institute. Washington, DC. October 2001. Available online at http://www.ifpri.org/sites/default/files/publications/fpr31.pdf.

PCFF (2002) *Farming and Food: A Sustainable Future*, Policy Commission on Farming and Food, chaired by Sir Donald Curry, London: Cabinet Office.

Pellizzoni, L. (2003) Knowledge, uncertainty and the transformation of the public sphere, *European Journal of Social Theory* 6 (3): 327–355.

People's Food Policy (2019) *A People's Food Policy: Transforming Our Food System*. Available Online at www.peoplesfoodpolicy.org.

Permaculture Association (2009) Knowledge base. Available online at http://www.permaculture.org.uk/.

Permaculture Association (2011a) The basics. Available online at http://www.permaculture.org.uk/.

Permaculture Association (2011b) *What is Permaculture*. Available online at http://www.permaculture.org.uk/.

Peterson, T.R. (1997) *Sharing the Earth: The Rhetoric of Sustainable Development*. Columbia, SC: Univ. of South Carolina Press.

Pink, S. (2015) *Doing Sensory Ethnography*. California: Sage Publications.

Ploeg, J.D. van der (1987) *The scientification of agricultural activities*. Wageningen LU.

Ploeg, J.D. van der (1993) Potatoes and Knowledge. In Hobart, M. (ed.) *An anthropological critique of development: The growth of Ignorance*. London and New York: Routledge: pp. 210–227.

Ploeg, J.D. van der, Renting, H., Brunori, G., Knickel, K. Mannion, J. Marsden, T.K., de Roest, K., Sevilla-Guzman, E. and Ventura, F. (2000) Rural development: from practices and policies towards theory. *Sociologia Ruralis* 40(4):391–408.

Polayni, K. (1957) *The Great Transformation: The Social and Economic Origins of Our Time*. Boston: Beacon Press.
Polanyi, K. (1977) *The Livelihood of Man*. New York: Academic.
Polanyi, M. (1967) *The Tacit Dimension*. New York: Anchor Books.
Policy Commission on Farming and Food (PCFF) (2002) *Farming and Food: A Sustainable Future*, Policy Commission on Farming and Food, chaired by Sir Donald Curry, London: Cabinet Office. http://archive.cabinetoffice.gov.uk/farming/pdf/PC%20Report2.pdf.
Ponte, S. (2009) From Fishery to Fork: Food Safety and Sustainability in the Knowledge-Based Bioeconomy, *Science as Culture* 18(4):483–495.
Poole, R. (1991) *Morality and Modernity*. London: Routledge.
Poole, R., Clarke, G., Clarke, D. (2002) Growth, concentration and regulation in European food retailing, *European Urban and Regional Studies* 9: 167–77.
Popper, K.R. (1972) *Objective Knowledge: an evolutionary approach*. Oxford: Clarendon Press.
Porter, T. (1995) *Trust in Numbers*. Princeton University Press.
Posey, D. (1999) Safeguarding traditional Resource Rights of Indigenous People. In *Ethnoecology: situated knowledge/located lives*. Virginia Nazarea: The University of Arizona Press. Tucson: pp. 217–230.
Poster, M. (1990) *The Mode of Information: Post-structuralism and Social Context*. Cambridge: Polity Press.
Pretty, J. (1995) Participatory learning for sustainable agriculture. *World Development* 23(8): 1247–63.
Pretty, J., Ball, A., Lang, T. and Morison, J. (2005) Farm costs and food miles: an assessment of the full cost of the weekly UK food basket, *Food Policy* 30(1):1–19.
Psarikidou, K. and Szerszynski, B. (2012a) Growing the Social: Alternative Agro-Food Networks and Social Sustainability in the Urban Ethical Foodscape, *Sustainability: Science, Practice and Policy* 8(1): 30–39.
Psarikidou, K. and Szerszynski, B. (2012b) The Moral Economy of Civic Food Networks in Manchester, *International Journal for the Sociology of Agriculture and Food* 19(3): 309–327.
Psarikidou, K., Kaloudis, H., Fielden, A. and Reynolds, C. (2019) Local Food Hubs in deprived areas: De-stigmatising Food Poverty? *Local Environment: The International Journal for Justice and Sustainability* 24(6): 525–538.
Raynolds, L. (2000) Re-embedding global agriculture: the international organic and fair trade movements. *Agriculture and Human Values* 17:297–309.
Raynolds, L.T. (2004) The Globalisation of Organic Agro-Food Networks, *World development*. 32(5):725–743.
Reich, R. (1991) *The Work of Nations: Preparing Ourselves for the 21st Century Capitalism*. London: Simon and Schuster.
Renting, H., Marsden, T. and Banks, J. (2003) Understanding alternative food networks: exploring the role of short food supply chains in rural development, *Environment and Planning A* 35(3): 393–411.

Reynolds, L. and Szerszynski, B. (2010) Contested agro-technological futures: the GMO and the construction of European space. In Robbins, P. and Huzair, F. (Eds.) *Transitioning the Life Sciences*, Springer.

Rigby, D. and Bown, S. (2007) Whatever happened to Organics? Food, Nature and the Market for 'Sustainable' Food, *Capitalism Nature Socialism* 18(3):81–102.

Rose, N. (2008) The Value of Life: Somatic Ethics and the Spirit of Biocapital, *Daedalus* 137(1): 36–48.

Royal Horticultural Society (2011) Gardening. Fleece. Available online at http://www.rhs.org.uk/.

Rudiger, K. and McVerry, A. (2007) *Exploiting Europe's Knowledge Potential: 'Good Work' or 'Could do better'. A Report prepared for the Knowledge Economy Programme November 2007.* The Work Foundation. Available online at http://www.agoratalentia.es/documentos/trabajadoresdeleconomiadelconocimiento.pdf.

Ryles, G. (1949) *The Concept of Mind*. London: Hutchinson.

Sage, C. (2003) Social embeddedness and relations of regard: alternative 'good food' networks in south-west Ireland. *Journal of Rural Studies* 19:47–60.

Schiellerup, P. (2005), *Identity and Environmental Governance: Institutional Change in Contemporary British Forestry Policy and Practice*, PhD, University College London.

Schumpeter, J. (1942) *Capitalism, Socialism, and Democracy*. New York: Harper.

Scott, J.C. (1998) *Seeing Like a State. How Certain Schemes to Improve Human Condition Have Failed*. Yale University Press.

Seale, C. (ed.) (2004) *Social Research Methods*. London: Routledge.

Seyfang, G. (2006) Ecological citizenship and sustainable consumption: examining local organic food networks, *Journal of Rural Studies* 22(4):383–395.

Shire, K. (2007) Gender and the Conceptualisation of the Knowledge Economy in Comparison. In Walby, S., Gottfried, H., Gottschall, K. and Osawa, M. (eds.) *Gendering the Knowledge Economy: Comparative Perspectives*. Palgrave Macmillan: pp. 51–78.

Sielbert, R., Larchewski, L. and Dosch, A. (2008) Knowledge dynamics in valorising local nature, *Sociologia Ruralis* 48(3): 223–240.

Sillitoe, P. (1998) The Development of Indigenous Knowledge: A New Applied Anthropology, *Current Anthropology* 39(2): 223–252.

Sillitoe, P. (2006) Introduction: Indigenous Knowledge in Development. *Anthropology in Action* 13(3): 1–12.

Simmel, G. (1990) *The Philosophy of Money* (edited by D.P. Frisby). London: Routledge.

Shiva, V. (1989) *Staying Alive: Women, Ecology and Development*. London: Zed Books.

Shiva, V. (1999) *Stolen Harvest: The Hijacking of the Global Food Supply*, Cambridge Massachusetts: South End Press.
Smith, A. (2006) Green niches in sustainable development: the case of organic food in the United Kingdom, *Environment and Planning C: Government and Policy* 24: 439–458.
Smith, E. and Marsden, T.K. (2004) Exploring the limits to growth in UK organics: beyond the statistical image, *Journal of Rural Studies* 20: 345–357.
Smith, K. (2000) What is the 'knowledge economy'? Knowledge-intensive industries and distributed knowledge bases. Paper presented to DRUID Summer Conference on The Learning Economy—Firms, Regions and Nation Specific Institutions. June 15–17, 2000.
Smith, K. (2002) What is the 'Knowledge Economy'? Knowledge Intensity and Distributed Knowledge Bases. UNU/INTEC Discussion paper. The United Nations University. Institute for New Technologies. Available online at http://eprints.utas.edu.au/1235/1/2002-6.pdf.
Soil Association (2011a) What is organic. Available online at http://www.soilassociation.org/.
Soil Association (2011b) Organic Standards. Available online at http://www.soilassociation.org/whatisorganic/organicstandards.
Soil Association (2011c) Soil Association Organic Standards: Farming and Growing. Available online at http://www.soilassociation.org/LinkClick.aspx?fileticket=l-LqUg6iIlo%3d&tabid=353.
Soil Association (2011) *What we do*. Available online at http://www.soilassociation.org/.
Sonnino, R. and Marsden, T.K. (2006) Beyond the Divide: Rethinking Relations Between Alternative and Conventional Food Networks in Europe, *Journal of Economic Geography* 6(2): 181–199.
Steel, C. (2009) *Hungry City: How food shapes our lives*. London: Vintage Books.
Strange, S. (1986) *Casino Capitalism*. Oxford: Blackwell.
Stuiver, M., Leeuwis, C. and van der Ploeg, J.D. (2004)The Power of Experience: Farmers' Knowledge and Sustainable Innovations in Agriculture. In J.S.C. Winskerke and J.D. van der Ploeg (eds.) *Seeds of Transition: Essays on novelty production, niches and regimes in agriculture*. The Netherlands: Koninkijke van Gorcum BV.
Sunder Rajan, K. (2006) *Biocapital: The Constitution of Postgenomic Life*. Durham and London: Duke University Press.
Swanborn, P. (2010) *Case Study Research: what, why and how?* London: Sage.
Szerszynski, B. (1993) Uncommon Ground: Moral Discourse, Foundationalism and the Environment Movement. Unpublished PhD Thesis, Lancaster University.
Szerszynski, B. (2005) *Nature, Technology and the Sacred*. Oxford: Blackwell.

Thacker, E. (2005) *The Global Genome: Biotechnology, Politics and Culture.* Cambridge, MA: MIT Press.

Thorpe, C. (2011) Artificial life on a dead planet. *Science as Culture*, in preparation.

Thanassoulis, J. (2009) *Supermarket Profitability Investigation.* University of Oxford, January 2009.

Toffler, A. (1980) *The third wave.* Bantam Books.

Torjusen, H., Lieblein, G., Vittersø, G. (2008) Learning, communicating and eating in local food systems: the case of organic box schemes in Denmark and Norway, *Local Environment* 13(3): 219–234.

Tovey, H. (2008) Introduction: Rural Sustainable Development in a Knowledge Society Era, *Sociologia Ruralis* 48(3):185–199.

Tovey, H. and Mooney, R. (2007) CORASON final report. Available online at http://www.corason.hu/download/final_report.pdf.

Trentmann, F. (2007) Before 'fair trade': empire, free trade, and the moral economies of food in the modern world, *Environment and Planning D: Society and Space* 25:1079–1102.

Tribe, K. (1978) *Land, Labour and Economic Discourse.* London: Routledge & Kegan Paul.

Tribe, K. (1981) *Genealogies of Capitalism.* London and Basingstoke, The Macmillan Press.

Turnbull, D. (1997) Reframing knowledge and other local knowledge traditions. *Futures*, 29(6): 551–562.

Turnbull, D. (2000) *Masons, Tricksters and Cartographers. Comparative Studies in the Sociology of Scientific and Indigenous Knowledge.* Harwood Academic.

Turnbull, D. and Verran, S. (1995) Science and Other Indigenous Knowledges in Sheila Jasanoff et al. (eds.), *Handbook of Science and Technology Studies*, London/Thousand Oaks, Sage (revised edition 2002): pp. 115–139.

United Nations (2005) *2005 World Summit Outcome*, resolution adopted by the General Assembly (A/RES/60/1), New York: United Nations.

Walby, S. (2007) Introduction: Theorising the Gendering of the Knowledge Economy: Comparative Approaches. In Walby, S., Gottfried, H., Gottschall, K. and Osawa, M. (eds.) *Gendering the Knowledge Economy: Comparative Perspectives.* Palgrave Macmillan: pp. 3–50.

Waldby, C. (2002) Stem Cell, Tissue Cultures and the Production of Biovalue, *Health* 6(3): 305–323.

Walby, S., Gottfried, H., Gotschall, K. and Osawa, M. (eds.) (2019) *Gendering the Knowledge Economy: Comparative Perspectives.* Palgrave MacMillan.

Wallace, H. (2010) Bioscience for Life? Who Decides what research is done in health and agriculture? Genewatch UK Report. March 2010.

Ward, S.V. (1992) *The Garden City: Past, Present and Future.* Oxford: Taylor and Francis.

Ward, N. (1994) *Farming on the treadmill: agricultural change and pesticide pollution.* Unpublished PhD Thesis, University College London.

Watson, T.J. (1994) *In Search of management: Culture, Chaos and Control in managerial World*. London: Routledge.

Watts, D.C.H., Ilbery, B. and Maye, D. (2005) Making reconnections in agro-food geography: alternative systems of food provision, *Progress in Human Geography* 29(1): 22–40.

Whatmore, S. and Thorne, L. (1997) Nourishing networks: alternative geographies of food. In Goodman, D. and M. Watts (Eds.) *Postindustrial Natures: Culture, Economy and Consumption of Food*. London, Routledge: pp. 287–304.

Whatmore, S., Stassart, P. and Renting, H. (2003) What's alternative about alternative food networks? *Environment and Planning A* 35: 389–391.

Williams, C. (2004) The Myth of Marketization: An Evaluation of the Persistence in Advanced Economies, *International Sociology*, 19: 437–449.

Wilsdon, J., Wynne, B. and Stilgoe, J. (2005) The Public Value of Science. Or how to ensure that Science really matters. *Demos* Available online at http://www.demos.co.uk/publications/publicvalueofscience. Retrieved May 25, 2009.

Winter, M. (2003) The policy impact of the foot and mouth epidemic, *Political Quarterly* 74(1): 47–56.

WIPO (2002) Information note on traditional knowledge, WIPO international forum on intellectual property and traditional knowledge: our identity, our future Muscat, January 21–22.

Wolf, E.R. (1969) *Peasant Wars of the Twentieth Century*. New York: Harper and Row.

World Commission on Environment and Development (1987) *Our Common Future*. Oxford: Oxford University Press.

Wynne, B. (2004) May the sheep safely graze? A reflexive view of the expert-lay knowledge divide. In Lash, S., Szerszynski, B. and Wynne, B. (eds.) *Risk, Environment and Modernity: towards a new ecology*. London: Sage publications.

Wynne, B. (2005) Risky Delusions: How GM Science has imagined—and provoked—its publics. In Taylor, I. and K. Barrett (Eds.) *Genetically Engineered Crops: Decision-making under Uncertainty*. Canada, UBC Press.

Wynne, B. (2007) Risky Delusions: Misunderstanding Science and Misperforming Publics in the GE Crops Issue. In I.E.P. Taylor (ed.) *Genetically Engineered Crops: Interim Policies, Uncertain Legislation*. Haworth Press.

Index[1]

A
Agricultural bio-technology, 32
Agriculture
  industrial, 5, 55–60
  organic, 4, 10, 60–61, 67n13, 80, 102–105, 108
Agrifood
  economy, 76, 127
  knowledge, 66, 67, 78, 79, 100, 111–113, 123–127, 135–138
  supply chains, 75, 77, 81, 117
  system, 115, 119, 121, 122, 124, 125, 127, 136
Agri-industrial complex, 56
Alternative, vii, 2–11, 35, 40, 41, 45, 58, 60–62, 66–68, 67n13, 74, 78–83, 86–93, 99, 102, 103, 106, 107, 110, 112–115, 117, 118, 120–128, 133, 135–139
  knowledge-based agro-food systems, 126

Alternative agro-food networks (AAFNs), vii, 1–12, 41, 45, 60, 63, 66–68, 73–93, 99–128, 133, 135–139
Animals, 24, 30, 57, 58, 81, 82, 103, 104, 105n2, 108, 109, 115, 118

B
Bell, Daniel, 19
Biodynamic, 80, 81, 87, 88, 103
Bioeconomy, 8, 10, 28–33, 36, 38, 39, 74, 76, 76n5, 77, 79, 90–93, 100, 127, 133
Biosciences, 30
Biotechnological revolution, 49
Biotechnology, 25, 28–33, 35, 36, 74, 76, 77
Blackler, F., 49, 50, 51n2, 113, 126, 134
Box scheme(s), 81, 84, 86, 88, 90, 105, 107, 114–118

---

[1] Note: Page numbers followed by 'n' refer to notes.

C

Capitalism
  casino, 37
  new face and phase of, 36
Capital(ist) accumulation, 21–23, 26, 29, 31, 33, 39, 48, 66, 128
Capitalocentric, 1, 9, 11, 12, 40, 100, 127, 128, 135–137, 139
Codification, 31, 39, 40, 106
Cognitive turn, 1
Collective
  action, 121
  knowing, 120
  learning, 120, 121
Commodification, 1, 2, 5, 7, 24, 26, 27, 30, 31, 33, 34, 36, 38–40, 64, 66, 67, 101, 127, 133, 136
Commodity, 1, 9, 21–23, 23n5, 26, 31, 34–36, 38–40, 47, 63, 67, 101, 127, 133
  fictitious, 22, 40
Common Agricultural Policy (CAP), 34, 80, 84
Communities, viii, 39, 49, 52n5, 53, 57, 65, 77, 80, 82, 84, 85, 87–90, 101, 109, 115–118, 120, 121, 124, 125, 134–136, 138
Communities of practice, 51, 52, 52n5, 65, 66, 106, 113, 125, 126, 134, 136
Community food growing, 90, 119
Community gardens, 82, 83, 92, 93, 101, 120
Community spaces, 120
Consumers, 2, 3, 27n9, 49, 60, 61, 77–84, 77n6, 87, 88, 93, 114, 116–121, 124, 125
Co-operatives, 81, 84, 86–89, 93, 107, 108, 113–116, 119, 120, 125

Co-production, viii, 32, 36, 46–48, 65, 121, 138
Creative destruction, 20, 20n4, 27, 36, 99, 126, 135
Crops, 2, 32, 33, 56, 57, 74, 75n2, 76, 77, 103, 104, 108, 108n4
Cumbria, 73, 83–90, 92, 102–108, 112, 113, 115, 117, 118, 121, 135

E

The economic, 46, 128, 136, 137
Environment(s), viii, 3, 4, 28, 55, 59, 62–64, 80, 81, 89, 91–93, 107, 108, 110, 123, 124
Experts, vii, 27, 58–62, 65, 66, 105, 107, 108, 110, 123–127, 134–138
  (un)certified, 58, 59, 66
Externalisation/internatisation, 51, 64, 122, 124, 125, 134

F

Farming
  farmers, 2, 3, 56–61, 63, 64, 65n11, 75, 79, 81, 81n9, 84, 84n12, 86–88, 102–108, 112–115, 117, 134
  farms, 2, 8, 33, 34, 56, 58, 74, 80, 87, 88, 92, 103–105, 107, 108, 114, 115, 117, 118, 135, 137
Farm shops, 81, 84, 86, 88, 92, 93, 113, 115, 117
Food
  fresh, 81–82
  local, viii, 4, 80, 88–90, 109, 111
Future, 3, 6, 7, 9, 10, 19, 24, 26, 27, 30, 33–40, 46, 48, 49, 57, 58, 86, 108, 113, 117, 128, 137

G

GeneWatch UK, viii, 33
Gibson-Graham, J.K., 11, 128, 137–139
Giddens, Anthony, 111, 116, 136
Growth, 1, 19–22, 21n4, 24, 26–30, 34, 38–40, 49, 74, 78, 82, 85, 101, 134, 135, 137–139

H

Haraway, Donna, 46, 55, 59, 62, 90, 91, 101, 126, 136
Hardt, M., 9, 21–24, 23n6, 63, 128, 133, 134
Health, 23n6, 25, 28, 30, 32, 77, 78, 81, 85, 86, 88, 89, 108, 115
Human/non-human, 1, 4, 22, 23, 29–32, 35, 39, 48, 55, 66, 89, 93, 104, 108–110, 123, 126, 128, 133, 136
Hybrid collectives, 55
Hybrid forums, 112
Hype, 32, 35–38

I

Iceberg economy, 11, 137
Imaginaries
 economic, 10, 37, 37n11, 48
 socio-technical, 38
 techno-scientific, 37, 38
Inclusion/inclusive, vii, 3, 9, 12, 60, 112, 121, 137–139
 knowledge economies, 137, 138
Information, 19, 21–26, 28, 30–32, 36, 46, 47, 50, 52, 79, 87, 92, 101, 118, 127, 128, 133
Information and Communication Technologies (ICTs)/IT, 9, 25, 27n9, 114, 126, 135, 138

Ingold, Tim, 55, 59, 117, 123, 124, 126, 136
Innovation
 agrifood, 8, 58, 66, 135, 137, 138
 potential, 99, 137–139
 radical, 60, 99, 126, 135
 techno-scientific, 32, 101

J

Jessop, Bob, viii, 1, 9, 10, 22, 27, 37, 37n11, 48

K

Knowing, 47, 48, 50–53, 58, 62, 64, 65, 67, 104, 106, 109, 110, 112, 120, 121, 123–126, 135, 137, 138
 agents, 117, 124, 126, 134–136, 138
Knowledge(s)
 agricultural knowledge, 33, 56, 58, 59, 61, 104, 106, 109, 111
 conversion, 51, 52, 125–127, 135
 economy, vii, 1–12, 19–41, 45–68, 73–93, 99–128, 134–139
 explicit, encoded and embrained, 10, 11, 22, 26, 45–54, 51n3, 52n5, 57–67, 73, 99, 106, 107, 112, 115, 122–126, 134
 hierarchies, 100, 106, 122, 124, 138
 immutable mobile/mutable immobile, 54, 55
 laboratory, 33, 34, 36
 lay, 46, 54, 58, 61, 62, 65n11, 112, 126
 local, 11, 46, 54, 57, 59, 60, 62, 99, 112, 123, 126, 135, 138
 new knowledge, vii, 12, 28, 31, 47–53, 51n3, 59, 62–65, 99, 111, 119, 123–128, 134, 137–139

Knowledge(s) (*cont.*)
  practical, 50, 105, 106, 111, 112, 126
  practices, 8, 55–57, 59, 60, 63–65, 79, 91, 93, 99, 103, 106, 107, 114, 128, 135, 136
  production, 2, 6, 8, 10, 20, 34, 41, 45, 47–64, 66–68, 100–128, 133–135
  scientific, 11, 20, 26, 28, 30, 33, 45–47, 50, 53–55, 58, 59, 61, 64, 90, 99–102, 104–107, 112, 115, 122, 123, 126, 127, 135
  sharing, 60, 111, 113, 126, 127, 136
  situated, 46, 55, 59, 100, 102, 106–108, 110, 112, 116, 118, 122, 123, 126, 135
  society, 6, 46
  tacit, embedded and embodied, 10, 11, 46–67, 51n3, 99, 105–108, 111–113, 116, 117, 120, 122–127, 134–138
  techno-scientific, 1, 26, 32, 34, 40, 53, 65, 101, 125, 126, 135
  traditional, 26, 46, 47, 53, 54, 56, 59, 59n10, 60, 64, 106, 107, 112, 122, 138
  transfer, 106, 120, 125
  workers, 26, 26n8, 27, 39, 41, 55, 63–65, 134–136
Knowledge-based bio-economy (KBBE), 6, 8, 10, 28–39, 63, 67–68, 127, 128, 136
Know who/why/how, 1–12, 40, 47–49, 51, 53, 61, 67, 80, 90, 101, 104, 106, 107, 110, 111, 113–118, 121, 123

L

Labour
  affective, 23, 23n6, 136
  agricultural, 66
  (im)material, 9, 21, 23, 23n5, 27, 35, 41, 63, 65, 66, 127, 128, 134
  naturally necessary labour time, 31, 35
  (un)productive, 22
Land, 21, 32, 54, 55, 57, 58, 63, 65, 65n11, 74, 74n1, 75, 75n2, 80, 82, 82n10, 83, 88, 89, 104, 105, 107, 108, 110, 111, 123, 124, 128
Latour, Bruno, 52, 54, 104, 110, 112, 123, 125, 134
Lazzarato, M., 9, 21, 23, 23n5, 128, 133
Learning
  collective, 52, 120
  by doing, 52, 107, 112, 134
  ongoing, 108, 111, 126
Life/bio
  as information, 31
  sciences, 29, 32, 36
Local authority, 85, 86, 89, 90, 119

M

Manchester, 73, 83–90, 92, 101, 103, 108, 109, 112, 115, 116, 118–121, 135
Market, 21n4, 25, 31, 57, 74, 75, 75n3, 80, 81, 81n9, 84, 84n12, 90, 92, 101, 102, 105, 108, 109, 114, 119
Marx, Karl, 19, 19n1, 20n3, 22, 29, 29n10, 34
Master economic narrative, 9, 27, 32, 48, 52, 134, 137
Merton, R. K., 9, 10, 36, 38, 48
Metabolic rift, 29

N

Nature, 2, 4, 20, 22, 28–35, 38, 39, 46, 47n1, 54–56, 61, 63–66, 77, 81, 83, 88, 93, 103–105, 107–109, 111, 112, 118, 123, 124, 127, 128, 133, 136

Negri, A., 9, 22–24, 23n6, 63, 134
Nonaka, I., 49, 51, 51n3, 64, 122, 126, 134
Northwest England, 83, 86, 100–121, 135

O

OECD, 1, 9, 24–26, 30, 32, 35, 49, 52, 73, 75n4
Off-farm/on-farm
  knowledge, 64, 114
  practices, 29, 56, 65, 66
Organic
  agriculture, 4, 10, 60–61, 67n13, 80, 102–105, 108
  certification, 105n2, 116
  farming, 60, 103, 104

P

Participation, 66, 93
Permaculture, 81, 89, 90, 103, 108–111
Plants, 22, 30, 33, 55, 56, 58, 64, 77, 81, 108–110, 108n4
Plural
  knowledges, 109, 125, 138
  pluralising knowledge economies, 137, 139
Polanyi, Michel, 47, 47n1, 48, 62, 102, 112, 122, 123
Policy, vii, 3, 9, 27, 32–34, 40, 134, 137, 139
  agendas, 1, 9, 12, 37, 40, 138, 139
Political act, 83, 120
Post-industrial economy, 1, 40
Prescriptive narrative, 28, 38, 39
Producers, 3, 59–61, 75n4, 77, 79–82, 84, 84n12, 87–89, 93, 114–118, 120, 124, 125

Productivity, 19, 21–24, 26, 27n9, 28, 30, 31, 33, 34, 55, 56, 63, 64, 66, 75, 101, 127, 128, 133, 134, 136–138
Profit(ability), 1, 22, 26, 28, 34–36, 38–40, 57, 64–66, 101, 127, 128, 136
Promise, 1, 10, 21, 31–33, 35–40, 75, 134, 137
Prosumers, 82
Proximity, 4, 52, 81, 84, 91, 114–116, 120, 125

R

Re-combination, 11, 49, 58, 64, 100, 124–126, 134, 135
Reconnection, 4, 81, 87, 88, 121
Regime
  accumulation regime, 27, 37, 48
  food regimes, 2, 3, 58
Research, vii, viii, 1, 2, 5–10, 12, 22, 25, 26, 30, 33, 34, 36, 37, 40, 62, 68, 73, 74, 76–79, 77n6, 86, 89–93, 100–102, 103n1, 105, 134, 137–139
  research agendas, 28, 32, 40
Re-skilling/de-skilling/up-skilling, 111, 114, 116, 118–121, 124, 125
Retailers, 2, 3, 77, 80, 82, 86, 87, 89, 114, 115, 117, 118, 120, 124, 125
Revolution
  green, 32, 55
Rural, 3, 6, 11, 73, 80, 84, 86, 87, 91, 99, 103, 112, 118, 135

S

Schumpeter, Joseph, 9, 20, 20–21n4
Science, 2, 20, 26, 28–30, 32–34, 36, 46–48, 54–59, 58n9, 61–63, 74, 75, 91, 100–112, 114, 122, 122n5, 123, 135, 138
  post-normal, 58

Scientific Knowledge, 11, 20, 26, 28, 30, 33, 35, 47, 50, 53–55, 58, 59, 61, 64, 90, 99–102, 104–107, 112, 115, 122, 123, 126, 135
Self-fulfilling prophecy, 10, 36, 38, 40, 48
Self-help, 87, 113
Situated knowledges, 59, 110, 116, 118, 123, 126, 135
Skills
  buying, 121
  communicative, 112, 115, 127, 135
  cooking, 115, 121
  marketing, 87, 113, 114, 124, 135
  organisational, 84, 115
  social, 51, 61, 113, 114, 116, 120, 121
Social cognitive spaces, 116, 117, 120, 121, 125, 126, 128, 134, 135
Social enterprises, 82, 87–89, 102, 104, 107, 109, 113, 115, 118, 121
Socialisation, 51, 134
Social relations, 19, 23, 23n5, 38, 39, 41, 52, 67, 119–121, 128, 136
Socio-material, 63, 64, 66, 90–93, 102, 104, 108, 111, 116, 118, 120, 123, 127, 134, 136
Socio-natural, 4, 124
Soil, 58, 63, 65n11, 80, 81, 103, 105, 107–110
Supermarkets, 78, 80, 81, 84, 84n12, 88, 114
Sustainability, 4, 6, 86, 87, 101, 117, 134, 137, 138
  agrifood, vii, 137, 139

Sustainable
  agriculture, 7, 34
  capital, 32

T
Technocentric, 1, 9, 12, 40, 137, 139
Truth, 28, 35–38, 37n11, 38n12, 40, 47, 58–60, 62, 91, 102

U
Urban, 11, 74, 83, 85, 86, 90, 91, 99, 101, 108, 110–112, 119, 120, 138
  food networks, 89

V
Value
  exchange, 36
  sign, 136
  surplus, 22, 23, 27, 31, 33–36, 38, 39, 127, 128, 136
  use, 36, 38, 39
Vision, 1, 6–8, 19, 24, 27, 30, 32–34, 37–41, 46, 48, 49, 51, 74, 79, 85, 101, 118, 133–135, 137, 139

W
Wellbeing, 85, 89, 118

Y
Yield, 75, 79, 110

The manufacturer's authorised representative in the EU is Springer Nature Customer Service Centre GmbH, Europaplatz 3, 69115 Heidelberg, Germany. If you have any concerns regarding our products, please contact ProductSafety@springernature.com

Printed and bound by CPI Group (UK) Ltd, Croydon, CR0 4YY

25/03/2026

02078205-0011